SpringerBriefs in Computer Science

For further volumes:
http://www.springer.com/series/10028

Cailian Chen • Shanying Zhu • Xinping Guan
Xuemin (Sherman) Shen

Wireless Sensor Networks

Distributed Consensus Estimation

 Springer

Cailian Chen
Department of Automation
Shanghai Jiao Tong University
Shanghai
China

Xinping Guan
Department of Automation
Shanghai Jiao Tong University
Shanghai
China

Shanying Zhu
Department of Automation
Shanghai Jiao Tong University
Shanghai
China

Xuemin (Sherman) Shen
Department of Electrical and Computer Engineering
University of Waterloo
Waterloo, Ontario
Canada

ISSN 2191-5768   ISSN 2191-5776 (electronic)
SpringerBriefs in Computer Science
ISBN 978-3-319-12378-3   ISBN 978-3-319-12379-0 (eBook)
DOI 10.1007/978-3-319-12379-0

Library of Congress Control Number: 2014953216

Springer Cham Heidelberg New York Dordrecht London

Printed on acid-free paper

Springer is part of Springer Science+Business Media (www.springer.com)

Preface

The increasing applications of wireless sensor networks (WSNs) witness the fact that the cooperative effort of sensor nodes can accomplish high-level tasks with sensing, data processing, and communication. Instead of sending the raw data to the fusion centers, sensor nodes execute distributed estimation for practical applications by locally carrying out simple computation and transmitting only the required and/or partially processed data. However, the network-wide information fusion capability and efficiency of the distributed estimation remain largely under-investigated. Moreover, the large-scale of WSNs imposes distinguished challenges on systematic analysis and scalable algorithm design to satisfy fundamental estimation criteria.

In this monograph, we focus on network-wide estimation (and tracking) capability of physical parameters from the system perspective. This problem is of great importance since fundamental guidance on design and deployment of WSNs is vital for practical applications. The metrics of network-wide convergence, unbiasedness, consistency, and optimality are discussed by considering network topology, distributed estimation algorithms, and consensus strategy. It reveals from systematic analysis that proper deployment of sensor nodes and a small number of low-cost relays (without sensing function) can speed up the information fusion and thus improve the estimation capability of WSNs. In Chap. 1, we introduce the spatial distribution of sensor nodes and basic scalable estimation algorithms for WSNs. Brief review of the existing works is given in Chap. 2 to show the collaborative and distributed processing of information with sensor observations and local communications. In Chap. 3, we exploit the consensus based estimation capability for a class of relay assisted sensor networks with asymmetric communication topology. By explicitly taking the functional heterogeneity between sensor nodes and relay nodes into account, a distributed consensus-based unbiased estimation (DCUE) algorithm is proposed. In Chap. 4, for the same relay assisted networks but with symmetric communication topology, we investigate the problem of filter design for mobile target tracking over WSNs. Allowed with process noise of the target and observation noise of sensor nodes, consensus-based distributed filters are designed for sensor nodes to estimate the states (e.g., position and speed) of mobile target. In Chap. 5, it is exploited on how to deploy sensor nodes and relays to satisfy the prescribed distributed estimation capability. A two-step algorithm is presented to meet the requirements of network connectivity and estimation performance. Finally, we draw conclusions and give a discussion on future work in Chap. 6.

This monograph is hopefully found to be helpful for graduate students and professionals in the fields of networking, computing, and control who are working on the research of topology analysis, sensor fusion, distributed computation and optimization, and especially of distributed estimation and control over WSNs.

We would like to thank the following colleagues and students for their valuable comments and suggestions on the monograph: Mr. Jianzhi Liu, Mr. Jianqiao Wang, Dr. Yiyin Wang and Dr. Bo Yang from Shanghai Jiao Tong University, China, Dr. Tom H. Luan from Deakin University, Ning Lu, Ning Zhang, Nan Cheng, Miao Wang, Dr. Hassan Omar and Khadige Abboud from Broadband Communications Research Group (BBCR) at the University of Waterloo, Canada, Xiang Zhang from University of Electronic Science and Technology of China, Dr. Juntao Gao from Xidian University, China, and Prof. Li Yu from Huazhong University of Science and Technology, China. Special thanks are also given to Susan Lagerstrom-Fife and Jennifer Malat from Springer Science+Business Media for their great helps and coordination throughout the publication process.

Shanghai, P. R. China   Cailian Chen
Singapore   Shanying Zhu
Shanghai, P. R. China   Xinping Guan
Waterloo, ON, Canada   Xuemin (Sherman) Shen
August 2014

Contents

Abbreviations

DCUE	Distributed consensus based unbiased estimation
DOCF	Distributed optimal consensus filter
DoE	Disagreement of estimate
HDCUE	Homogeneous DCUE algorithm
HWSN	Heterogeneous wireless sensor network
KCF	Kalman-consensus filter
LMS	Least-mean-square
MSD	Mean-square deviation
MSE	Mean square error
RMSE	Root-mean-square error
RN	Relay node
SCC	Strongly connected component
SN	Sensor node
SNR	Signal-to-noise ratio
WCC	Weakly connected component
WSN	Wireless sensor network

Chapter 1
Introduction

Wireless sensor networks (WSNs) [1] are massively distributed systems for sensing and processing of spatially dense data. They are composed of large number of nodes deployed in harsh environments to execute challenging tasks including security/surveillance, environmental monitoring, health monitoring, industrial automation, and disaster management, etc. Although the nodes only have limited resources, complicated tasks such as distributed detection and estimation [2] can be accomplished via nodes' cooperation. The main argument is that a distributed sensor network can leverage its performance by aggregating information gathered by individual nodes, which is known as information fusion. The primary goal of sensor fusion is to process and progressively refine information from multiple nodes to eventually obtain situation awareness.

However, there is a gap between the wealth of the captured information and the understanding of the physical situation. The applications in various areas arise a common problem: for a given environment, how to estimate the spatial and temporal distribution of physical parameters of interest based on local node observations, which are probably disrupted by noise, in a distributed manner by means of simple computations and local data exchange. Thus, distributed estimation is a key process to bridge the gap by locally carrying out computations and transmitting only the required and/or partially processed data over WSNs.

1.1 Motivation of Distributed Consensus Estimation

WSNs have attracted increasing attention recently due to their wide applications. Significant advances have been made for communication, signal processing, routing, and sensor deployment or selection, etc. However, in many application scenarios, a large number of low cost sensors are deployed in order to sense or monitor the environments or equipments. This introduces a straightforward question: how to acquire the accurate and real-time status of the monitored fields or objects? To answer this question, we resort to the study of estimation. Distributed consensus estimation is the most widely adopted method in WSNs.

© The Author(s) 2014
C. Chen et al., *Wireless Sensor Networks*,
SpringerBriefs in Computer Science, DOI 10.1007/978-3-319-12379-0_1

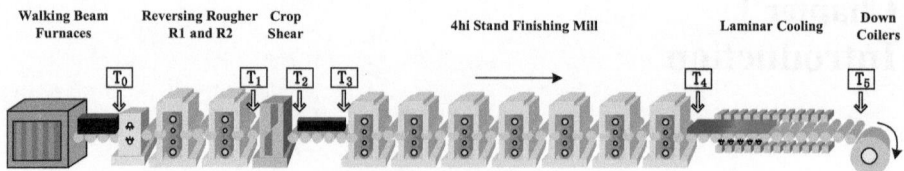

Fig. 1.1 A schematic view of hot rolling production line

There are two main reasons to apply the distributed consensus estimation in WSNs. The first is due to the unique structures of WSNs. Since the nodes often have capabilities of sensing, communication, and computation, they can self-organize into a network. There is no infrastructure to fuse all the information collected by sensor nodes. Moreover, even there is a fusion center, distributed estimation is still a better choice. In distributed estimation, each node does not need to send its data to a fusion center, thus it also does not need to establish and maintain a routing table for data packets as well. This can reduce the communication cost between nodes and improve the robustness of the network to node failures. This is extremely important if the communication topology is time-varying.

The second is due to the resource limitation of nodes. On one hand, due to the limited sensing capability, a node can only partly observe the global status of the target. The nodes need a cooperative way to communicate with others to compute the global estimation. Consensus algorithms are effective tools to exchange information among the network. The effectiveness of consensus algorithms have been proven for performing network-wide distributed tasks. On the other hand, due to the power constraints and attenuation of radio propagation in wireless channels, it can only ensure reliable communications within a short distance. In distributed consensus algorithms, the nodes only need to exchange local information with their neighbors. Each node carries out local estimate and eventually obtains the accurate estimates of the global parameters.

Distributed consensus estimation algorithms play an important part in an amount of application scenarios. A typical application scenario is quality inspection in industrial automation. A wealth of process data can be obtained through sensor nodes by using industrial wireless technology. However, the temperature, humidity, pressure, and vibration intensity measurements are only raw data. The gap between the captured raw data and the extracted useful information can be bridged by distributed consensus estimation. Through local computation and communication, estimates of all the nodes are ensured to converge to true values of the physical process. Fig. 1.1 shows an application example: the hot rolling process which is composed of several stages. The temperature at the entrance of rouging mill stage is vital for the product quality in the whole process. In more details, slab temperature will critically influence the calculation of the rolling force and the metallurgical and mechanical properties of the final products. Traditional, temperature detection approaches only sense the temperature of isolate points in the slab, which is not precise. In this case, a distributed consensus scheme can be applied to estimate the temperature distribution in the hot rolling process. As a result, the quality of the final slab products will be improved significantly by a more precise estimation of the temperature.

1.2   Overview of Distributed Consensus Estimation

Distributed consensus estimation algorithms in general can be categorized into three groups: observation-only consensus, observation+innovation consensus, and consensus-based filters for dynamic targets. Starting with the general observation model, we will specify the characteristic of each group. In WSNs, each node can obtain a measurement of the global target. Assume the estimated global target is x, and the general observation model is described as $z = Hx + w$ with H being the observation matrix. The classification is given by taking into account the following two considerations: (1) dynamic information of the target, and (2) the strategy used in the estimation process. Specifically, observation and estimation are two actions in the whole estimation process. In the observation-only consensus, observation and estimation are two independent actions. The nodes firstly obtain the measurements and then cooperatively estimate the target in a distributed manner until consensus is achieved. However, the observation+innovation strategy interwinds the observation and estimation actions. At each step of estimation action, each node takes the innovation (a new measurement value through the observation action) and utilizes the innovation to modify the whole estimation process. The observation+innovation consensus and the consensus-based filter for dynamic target both use the latter strategy. The difference between them is whether the target is static or dynamic. In general, observation-only consensus algorithms are often adopted to estimate the static target. Hence, observation-only consensus is not an appropriate option to estimate a dynamic target.

1.3   Distributed Observation and Consensus Strategy

1.3.1   Observation-Only Consensus

Observation-only consensus is a class of distributed consensus algorithms composed of two independent actions. The nodes take the observation action and utilize the measurements to process the estimation action. According to the approaches adopted in the estimation action, it can be further divided into two subclasses: average consensus and in-network regression consensus. According to the observation model mentioned in the previous section, we can further understand the difference between them. The goal of estimation is to find out the accurate value of x from the observation z. Notice that observation-only consensus works only if the target x is stationary. In this case, the regression analysis method is an appropriate and effective choice which is employed in the in-network consensus. However, there are also some special situations where either the observation matrix satisfies $H = I$ or the state of z is focused instead. Spectrum sensing in cognitive radio networks is a typical case. Radio signal strength at the locations of sensor nodes are to be estimated as well as the power of source nodes. Average consensus is a class of algorithms designed for these special application scenarios. In fact, this makes the research emphases of in-network regression consensus and average consensus quite different.

1.3.2   Observation+Innovation Consensus Strategy

Observation+innovation consensus is a kind of linear iterative algorithms that interwind observation action and estimation action. There are three reasons leading to the combination of the two actions. First, the parameters to be estimated always fluctuate with the varying environment. New observations can provide more information about the latest states of the parameters to improve the estimation accuracy. Second, observation-only consensus requires the time scale of communication which is much smaller than that of observation [3]. However, in many circumstances, communication cost between nodes is high. And the observation and communication rates of nodes are always the same. Third, due to the limited sensing capability, a single node can only observe a part of the whole parameter vector. Since we aim at estimating the whole vector at every node, average consensus is not appropriate any more. In fact, the observation matrix H of observation+innovation consensus also has a special form. The special form means that each node can only partly observe the parameter vector. Convergence, consensus, estimation error, and the rate of convergence are metrics to evaluate the observation+innovation consensus [4].

Chapter 2
Distributed Consensus Estimation of Wireless Sensor Networks

Recently, consensus based distributed estimation has attracted considerable attention from various fields to estimate deterministic parameters and track time-varying ones. In this chapter, the state-of-the-art of distributed consensus estimation is discussed.

2.1 Consensus Based Distributed Parameter Estimation

2.1.1 Average Consensus

Average consensus develops with multi-agent systems where consensus is a vital aspect in coordination and cooperation [5]. It is a linear iteration scheme where each node updates its value as a linear weighted combination of the values received from neighbors and its own. Consensus can be guaranteed by appropriately designing the weights used in the linear schemes. However, the characteristics of WSNs introduce several challenges for average consensus as summarized below: (1) The nodes are always supplied by portable batteries whose energies are also constrained by limited physical sizes. As a result, efficiency is a vital aspect in average consensus. (2) Nodes need to exchange messages through unreliable wireless communication. The unreliability can introduce noise, dynamic topology, time delays, and other problems. How to obtain robust estimation is another issue should be concerned with. (3) The nodes are easy to be compromised and the security problem in estimation is also important.

Plenty of research work have been reported to tackle these challenges and they are classified in Table 2.1 . On the aspect of efficiency, designing protocols for reaching consensus with fast convergence rate is a choice. Matrix optimization is utilized to design the weight coefficients in [6]. The increased convergence time based on matrix optimization is limited by network connectivity. It can be slowed down even if the weights are optimized. A local prediction component is added to the update protocol in [7] and [8] which propose theoretical analysis to demonstrate the improvement of the convergence time. Lower bounds for iteration steps in average consensus and a minimum-time consensus scheme are also proposed in [9]. [10] couples the

© The Author(s) 2014
C. Chen et al., *Wireless Sensor Networks*,
SpringerBriefs in Computer Science, DOI 10.1007/978-3-319-12379-0_2

Table 2.1 Average consensus

Efficiency	
Fast convergence by designing weights	[6]
Fast convergence by a prediction component	[7, 8]
Achieving consensus with minimum iterations	[9]
Removing the computation of maximum degree	[10]
Transmitting with bounded peak power	[11]
Quantized communication data with time-varying topologies	[12]
A low complexity quantizer and refined quantization	[13]
Robustness	
Finding the weights causing least-mean-square deviation with channel noises	[14]
Convergence property under imperfect communications	[15]
Coupling consideration of channel noise and convergence rate	[16]
Convergence under Markovian random graphs	[17]
Average consensus with random topologies and noisy channels	[18]
Average consensus with asynchronous communications between sensors	[19]
Weighted average on directed graphs	[20]
Consensus over directed graphs with quantized communication	[21]
Directed networks with distributed time delays	[22]
Cyber-Security	
Analysis secure consensus through a system theoretic framework	[23, 24]
Considering two types of outlier attackers	[24]
Secure average consensus algorithms in spectrum sensing	[25, 26]
Privacy preserving consensus	[27]
Secure average consensus-based time synchronization protocol	[28]

computation of the consensus value and the estimation of Laplacian matrix that can remove the computation process of the maximum degree of the network. Notice that the main power consumption is in the communication of a node. Therefore, another way to save the energy is to use quantized communication data. A nonlinear average consensus scheme with bounded peak power is proposed in [11]. Every node proceeds a prior stage to map the data through a bounded function in order to bound the transmit power. A uniform quantizer with constant step size and a communication feedback component are introduced to deal with the sensor saturation and time-varying topologies in [12]. The correlation between the exchanged values during the consensus process is exploited in [13]. It results in a low complexity quantizer and refined quantization during the convergence process.

There are also some methods to address the second challenge. Channel noise is inevitable in WSNs. Thus, many standard consensus algorithms under perfect communication may fail to converge as observed in [14]. A solution is then provided to find the best edge weights resulting in optimal estimation. In [15], authors focus on the imperfect communications and prove the convergence property under some perturbation models of exchanged data between nodes. A scheme considering both the channel noise and convergence rate is proposed in [16]. Typical WSNs also suffer from link failures, packet drops, and node failures, which results in switching topology, time delays, and other problems. [17] shows some convergence results under Markovian random graphs using the theory of Markovian jump linear systems. Average consensus with random topologies and noisy channels are investigated in [18]. Two algorithms called A-ND and A-NC are proposed to address the trade-off between bias and variance caused by link failures and noisy channels. All the average consensus algorithms require clock synchronization which is hard to achieve. Asynchronous average consensus algorithms are appropriate to tackle this problem. It is known that the necessary condition for all sensors converge to the average value is that the sum value remains the same. [19] proposes an implementation that guarantees the necessary condition in spite of asynchronous communications between nodes. Weighted average consensus takes node measurement accuracy and environmental conditions into consideration which makes the estimation more accurate and reliable. Authors in [20] modify the existing weighted average consensus algorithms to remove the requirement of bidirectional communication between neighbors. As a result, the modified algorithm can work under directed graphs. The problem of reaching consensus of a general unbalanced directed network under limited information communication is addressed in [21]. Directed networks with distributed time delays are investigated in [22]. Single and multiple time delays are investigated, respectively.

Cyber-Security is another aspect that matters in distributed average consensus. And secure average consensus algorithms have been more and more important with the wide application of distributed average consensus protocols. It aims at ensuring trustworthy computation in linear iterations in the presence of malicious inner sensors or outer intrusions. References [23, 24] model misbehavior as unknown and unmeasurable inputs and address the detection and identification problem through an unknown-input system theoretic framework. Two types of adversarial outer attacks are considered in [24]. The adversary is either able to break a number of links or add noise on the values of the nodes. Both attacks are analyzed by optimal control theory. References [25, 26] apply the secure average consensus algorithms in spectrum sensing. The secure schemes can adaptively adjust the weights of neighbors and gradually isolate the malicious nodes. The adaptive threshold is also able to mitigate the misbehaviors of inside nodes. A PPAC algorithm is proposed to guarantee the privacy of the initial values while ensure the whole network converge to the exact average in [27]. The key point is to add and subtract random noises to the iterative process. Some theoretical analyses are also given in [27]. A secure average consensus-based time synchronization protocol is proposed in [28].

Table 2.2 In-network regression consensus

Consensus-based D-LS Using ADMM	[29]
DiCE: introducing new consensus constraints to reduce exchanged message	[30]
Fast-DiCE: fast convergence by using Nesterov's optimal gradient descend method	[31]
Consensus-based D-LS with quantization and communication noise	[32]
Consensus-based D-TLS	[34]
IPI-based D-TLS: reduced computational complexity	[35]
Two stage consensus-based solution for L norm regularization	[36]
Three stage solution with low complexity and memory requirement	[37]
PSSE: iteratively exclude abnormal values	[38]
Consensus-based framework from both attacker and defender aspects	[39]

2.1.2 In-Network Regression Consensus

As discussed in the previous subsection, we focus on the state of observation z or the observation matrix that is equal to the identity matrix I in average consensus. However, in many application scenarios, the state of the original target x matters. The observation matrix H often has a more general form. These problems are called linear inverse problems, and in-network regression consensus is a class of algorithms to solve them. Although in-network regression consensus is a subclass of observation-only consensus, it employs regression analysis methods like maximum likelihood and least squares estimation, which is different from average consensus. The difference leads to different research emphases. In in-network regression consensus, we always formulate the estimation of the target into a convex minimization problem which exhibits a separable structure. Using the separable characteristic of the problem, consensus-based distributed solutions are exploited. Despite of this basic formulation which directly uses the regression analysis methods, there are also algorithms considering more limitations that introduce regularization into the convex problem. Typical applications of in-network regression consensus include distributed spectrum sensing, distributed field estimation, distributed target localization and state estimation in smart grid, etc. References on in-network regression consensus are listed in Table 2.2. Reference [29] adopts the least squares (LS) technique to formulate the convex problem. By introducing the consensus constraints and following the method called alternating direction method of multiplier (ADMM), a distributed consensus algorithm is proposed. Considering new consensus constraints, a new algorithm called DiCE which can reduce the exchange messages between neighboring nodes is proposed in [30]. A Fast-DiCE that takes the advantage of Nesterov's optimal gradient descend method is then presented in [31]. However, these algorithms do not consider the communication noises and link failures which are unavoidable in WSNs. In [32], authors introduce a distributed consensus scheme for an LS problem which guarantees the convergence even in the presence of quantization or communication noise. Reference [33] investigates the performance of the algorithm when there are erroneous links between neighboring nodes. A scheme is also proposed in order to mitigate the influences and ensure satisfactory overall performance. A distributed TLS (D-TLS) is proposed in [34] to tackle the situation where the observation matrix H is also noisy. To reduce the large computational complexity caused by the process of eigenvalue decomposition in each step, a modified D-TLS called IPI-based D-TLS is proposed in [35]. Sometimes $\mathcal{L}$ norm regularization is added into the convex problem in order to improve the estimation accuracy or obtain stable solutions. This idea leads to the $\mathcal{L}$ norm recovery methods widely applied in compressed sensing, smart grid, field estimation, and other situations. Distributed consensus solutions are often chosen to solve the recovery problems. Basically, the introduction of the $\mathcal{L}$ norm regularization is dependent on the sparsity of the state to be estimated. [36] proposes a two stage algorithm to solve the $\mathcal{L}_1$ norm recovery problem. A model-robust adaptation is also adopted to control the approximation error caused by spatial quantization. An iterative thresholding and input driven consensus-based three-step method appears in [37] with low complexity and memory requirement. In order to obtain robust power state estimation, [38] proposes a distributed PSSE estimator based on ADMM to iteratively exclude the abnormal values. Sparse attack construction and state estimation are exploited in [39]. A distributed framework for both aspects are considered at the same time followed by corresponding distributed consensus algorithms.

2.1.3   Observation+Innovation Consensus

Considering the fluctuation of the deterministic parameters to be estimated and the timescales of communication and observation, the mentioned two classes of algorithms are not suitable. Observation+innovation consensus interwinds observation and estimation to tackle the problem. The estimation accuracy can be improved by introducing new observations during estimation process. And the observation matrix H also has a special form with some diagonal entries being zeros. Convergence, consensus, estimation error, and the rate of convergence rate are the important metrics to evaluate the observation+innovation consensus algorithms. They are enumerated in Table 2.3. Reference [4] provides an observation+innovation consensus algorithm for a deterministic target to remove the requirements of local observability. [40] proves the bounded estimation error of the algorithm and quantify the trade-off between connectivity, observability, and stability. Reference [41] gives a bound on the mean square of the convergence rate and studies the behavior of the algorithm with the measurements fading with time. The nonlinear observation models and noisy communication links are considered with theoretical analysis. Reference [42] addresses the problems of random link failures, stochastic communication noises, and Markovian switching topologies. Both the mean square and almost sure convergence are established. Quantization errors, successive packet dropouts, and randomly varying nonlinearities of the target are considered together in [43]. For non-Gaussian observations, there is a threshold of network degree of connectivity. If it is below

Table 2.3 Observation+innovation consensus

Removing the requirements of local observability	[4]
Quantifying the trade-off between connectivity, observability, and stability	[40]
A bound on the mean square of the convergence rate with measurement fading	[41]
Nonlinear observation models and noisy communication links	
Random link failures, stochastic communication noises, and Markovian switching topologies	[42]
Quantization errors, successive packet dropouts, and randomly varying nonlinearities	[43]
Analysing the gap between distributed algorithm and corresponding central algorithm	[3]
Applications of observation+innovation consensus	[44, 45]
Considering heterogeneous sensor networks	[46]

the threshold, a gap between distributed algorithm and its corresponding central algorithm appears. The conclusions can be found in [3]. Applications of observation+innovation consensus algorithms for economic dispatch in power systems and for wide area monitoring systems are described in [44] and [45], respectively. Notice that all the estimation models are homogeneous in the previous part, which means the nodes are identical in the network. However, heterogeneous sensor networks introduce different kinds of nodes in order to prolong the life of networks. The interesting work of applying the observation+innovation consensus in heterogeneous sensor networks is firstly addressed in [46].

2.2   Consensus Based Distributed Tracking

Estimation and tracking of dynamic targets is one of the main objectives of WSNs. However, the previous three classes of distributed consensus algorithms are not suitable for tracking dynamic targets. Although centralized filters like Kalman filters, particle filters can track the dynamical processes, they are not implementable in distributed WSNs. To solve the estimation problem in WSNs, a lot of distributed versions have been proposed as summarized in Table 2.4. On the aspect of distributed Kalman filters, Olfati-Saber first introduces a consensus-based Kalman filter inspired by the consensus strategy in [47]. The filter consists two stages: a Kalman like measurement update and an inserted consensus term to eliminate the disagreements of sensors. A further study of the optimality and stability performance of the algorithm is then examined in [48]. An alternative consensus-based Kalman filter is proposed in [49] with the investigation of the correlation between Kalman gain and the consensus matrix. Some parameters design guides are also given in the literature in order to minimize the estimation error. However, it is far from optimal in Kalman-consensus filter (KCF) because of the correlation between local estimates. Furthermore, it is hard to exactly determine the correlation that causes the nonoptimality. An adaptive consensus-based Kalman filter is proposed in [50]. By

Table 2.4 Consensus based filters for dynamic targets

Consensus-based distributed Kalman filter	[47]
Further study of the optimality and and stability performance	[48]
Investigating correlation between Kalman gain and the consensus matrix	[49]
An adaptive consensus-based Kalman filter	[50]
Information consensus-based filter	[51, 52]
Further improving performance by designing consensus weights	[53]
Considering the network induced delays and dropouts	[54]
Robust estimator addressing uncertain channels	[55]
Quantised communications and random sensor failures	[56]
Event-driven transmission schemes	[57, 58]
Distributed optimal consensus filter for heterogeneous networks	[46]
Considering a nonlinear system model	[59]
Distributed consensus-based particle filters	[60, 61]
Distributed particle filter for nonlinear tracking	[62]

adding extra exchanged information between nodes indicating whether or not a node observes the target, the algorithm can improve the estimation accuracy compared with KCF. Other techniques resorting to information filter have been developed in [51, 52] which give insight into the influence of the correlation. Based on the information consensus-based filter, a scheme designing the consensus weights to further improve the performance is presented in [53]. In practical applications, there are often network-induced phenomena, such as delays and packets dropouts. A scheme based on local Luenberger-like observers is proposed in [54] to address the network induced delays and dropouts. Considering uncertain channels, a robust estimator with adaptive channel estimator is presented in [55]. WSNs also suffer from power constraints in practical situations which makes the energy consumption problem important. Authors in [56] adopt the probabilistic strategy to reduce the energy consumption. Alternatives such as event-driven transmission schemes are provided in [57, 58]. Each node transmits a new data only when a predefined event happens, which can significantly reduce the transmission power. Distributed optimal consensus filter appropriate for heterogeneous sensor networks can be found in [46]. [59] extends the linear consensus-based Kalman filter to a nonlinear system model. In addition to consensus-based Kalman filters, there are some other approaches designed to reach consensus. Distributed consensus-based particle filters are developed in [60] and [61]. Both of them consist of two major steps with the difference being whether average consensus or support vector machine is used at the first step. To deal with the nonlinear systems, a corresponding unscented particle filter is proposed in [62].

Chapter 3
Consensus-Based Distributed Parameter Estimation with Asymmetric Communications

This chapter focuses on exploiting the sensor fusion capability for situation monitoring applications over a kind of relay assisted sensor networks consisting of multiple kinds of SNs and RNs. The SNs implement assimilation of new measurement and cooperation with other nodes. While for RNs, the main role is to aggregate their neighboring data. Moreover, SNs have different sensing modalities, which can only measure a part of the target parameter vector for situation monitoring. We propose a distributed consensus-based unbiased estimation (DCUE) algorithm for this kind of sensor network. Different from existing algorithms, the DCUE algorithm explicitly takes the heterogeneity of responsibilities between SNs and RNs into account. By using algebraic graph theory in conjunction with Markov chain approach, we demonstrate how the distributed estimation method can be transformed to circumvent the challenges arisen from the heterogeneity. We analyze the performance of asymptotic unbiasedness and consistency of the DCUE algorithm in the presence of asymmetric communication, i.e., when a node can receive information from another node but not vice versa. Furthermore, a quantitative bound on the rate of convergence is established. Finally, simulation results are provided to validate the effectiveness of the DCUE algorithm. It is also demonstrated that the presence of RNs does contribute to the estimation accuracy and convergence rate compared with the homogeneous networks.

3.1 Introduction

Recently, estimation in distributed sensor networks has attracted much attention from different communities, to name a few, [63–66]. However, most previous research results focus on networks with all nodes possessing identical responsibility in the sensing, processing, and communicating. One drawback of such networks is their limited performance [67]. An interesting question is whether it is better to use a combination of different nodes, e.g., a large number of low-end nodes and a few high-quality nodes, leading to the so-called heterogenous network. Several works have been presented to address the improvements of the heterogeneity in prolonging

© The Author(s) 2014   13
C. Chen et al., *Wireless Sensor Networks*,
SpringerBriefs in Computer Science, DOI 10.1007/978-3-319-12379-0_3

network life and balancing energy consumption [68, 69]. On the other hand, for typical applications of sensor fusion, it is general that heterogeneous nodes would coexist in the network with different signal characteristics. In such cases, it may not be possible to mix sensory data from different modalities.

Taking these observations into consideration, we use a kind of relay assisted sensor networks with SNs and RNs to estimate the parameters of a target. SNs are heterogeneous with widely varying signal characteristics, which are responsible for data fusion and do not take the traffic relaying as a routine function. Typical applications of such kind of networks include large-scale field monitoring, where several groups of SNs are clustered geographically. In order to prolong the lifetime of SNs, a popular approach is to deploy some RNs to connect these separate groups such that the whole network is connected, while meeting certain network specifications [70–72]. In this case, RNs may comprise of wireless transceivers but no sensors, mainly accounting for the aggregation function. Whether or not and how the heterogeneity of SNs and RNs can improve the network-wide estimation capability is unknown till now. In sensor networks, communication between certain pairs of nodes could be asymmetric, i.e., *when a node can receive information from another node but not vice versa*. In this chapter, we aim to propose a distributed estimation algorithm based on consensus strategy, which is compatible with the relay assisted design of the network. Our purpose is to design the algorithm such that all the SNs can obtain reliable estimates of the situation parameters without knowing all the parameters. This chapter is based on [74]. The main contributions are summarized as follows:

- A distributed consensus-based estimation (DCUE) algorithm is proposed for solving the estimation problem in asymmetric relay assisted sensor networks. For SNs, the algorithm combines an innovation term (*assimilation of new measurement*) and a local consensus strategy (*local averaging among nodes*). While for RNs, only aggregation operation is performed. We appropriately introduce a graph transformation on the communication graph to tackle the cycles consisting of RNs in characterizing the relaying feature of RNs, followed by a Markov chain based characterization.
- We analyze the asymptotic unbiasedness and consistency of the DCUE algorithm. In the presence of asymmetric communication and coexistence of SNs and RNs, the original system is hard to analyze. We propose an appropriate transformation on the original system and adopt some novel techniques to put the analysis in a tractable framework. Unlike [65, 73, 74], our method does not require the assumptions of bidirectional communication. It is applicable for the general asymmetric case.
- We provide a quantitative bound on the rate of convergence of the DCUE algorithm in the almost sure sense. Moreover, the comparison of convergence rate of a variant of the consensus algorithm derived from the DCUE algorithm with its homogeneous counterpart is given. We establish an upper and a lower bound of the gap between these two consensus algorithms for the case of bidirectional communications. It is found that the insertion of RNs indeed improves the convergence rate, provided an appropriate deployment strategy is adopted.

Notation: $\mathbb{R}^n$ is the n-dimensional Euclidean space with the spectral norm $\|\cdot\|$. $\mathbf{1}$ is the vector of ones, $\mathbf{0}$ stands for a vector or a matrix of zeros, I denotes the identity matrix and diag$\{\cdot\}$ represents the diagonal block matrix. The smallest, second smallest and largest eigenvalues of a symmetric matrix are denoted by $\lambda_{\min}(\cdot)$, $\lambda_2(\cdot)$ and $\lambda_{\max}(\cdot)$, respectively and its trace by tr$(\cdot)$. We denote the Kronecker product of two matrices A and B by $A \otimes B$.

3.2   Problem Formulation

3.2.1   Network Model

Consider the problem of estimating a vector-valued situation parameter $\theta \in \mathbb{R}^J$ in a monitored field. The field is deployed with a sensor network composed of two types of nodes: SNs and RNs, denoted by $\mathcal{I}_S := \{1, 2, \ldots, M\}$ and $\mathcal{I}_R := \{M+1, M+2, \ldots, N\}$, respectively. Each SN can sense the ambient parameter θ but only a subset, and the measurement of SN i is given by

$$y_i(t) = H_i \theta + w_i(t), \ \forall i \in \mathcal{I}_S, \tag{3.1}$$

where $H_i \in \mathbb{R}^{J_i \times J}$ is the observation matrix and $w_i(t)$ is white Gaussian noise. It is assumed that $w(t) = [w_1^T(t), w_2^T(t), \ldots, w_M^T(t)]^T$ has bounded covariance and is independent of t. The observation noises may be correlated across nodes at each time t. The spatial correlation of the observation noise makes our formulation more practical. Note that generally $J_i \leq J$, which means that each SN can only provide limited observations and θ is not observable at each SN. As for RNs, they can not measure θ due to hardware limitations, and their main role is to relay the data for SNs.

We assume that each node i transmits at a constant power level P_{Ti} to node j over wireless channel. Assuming that proper scheduling is executed among vicinity nodes so that the concurrent transmission is avoided and co-channel interference is negligible, the data from node i can be successfully received by node j if

$$\frac{P_{Ti}}{\mathbb{N} d_{ij}^{\eta}} \geq \beta, \tag{3.2}$$

where $\mathbb{N}$ represents the ambient noise power level, d_{ij} is the distance between nodes i and j, and η is the path loss exponent (typically, $2 \leq \eta \leq 6$). Equation (3.2) models a minimum signal-to-noise ratio of β necessary for successful receptions [67]. Thus, the node j should be within a distance of $r_i := \sqrt[\eta]{P_{Ti}/(\mathbb{N}\beta)}$ from the transmitter i in order for reliable communication. We call the nodes in $\mathcal{N}_i = \{j \in \mathcal{V} : d_{ij} \leq r_i\}$ the *neighbors* of transmitter i, and $\mathcal{N}_i^S = \mathcal{N}_i \cap \mathcal{I}_S$, $\mathcal{N}_i^R = \mathcal{N}_i \cap \mathcal{I}_R$ the *SN neighbors* and *RN neighbors*, respectively.

3.2.2 Concepts From Graph Theory

Due to the asymmetric communication links between nodes, we can model the communication network as a *weighted directed graph* $\mathcal{G} = (\mathcal{V}, \mathcal{E}, \mathcal{A})$, where $\mathcal{V} = \mathcal{I}_S \cup \mathcal{I}_R$ is the set of SNs and RNs; $\mathcal{E} \subset \mathcal{V} \times \mathcal{V}$ is the set of communication links; $\mathcal{A} = [a_{ij}]$ is the weight matrix associated with $\mathcal{E}$, meaning that $a_{ij} > 0 \Leftrightarrow (j, i) \in \mathcal{E}$. If $a_{ij} > 0 \Leftrightarrow a_{ji} > 0$, then it is called an undirected graph. The directed edge $(i, j) \in \mathcal{E}$ means that the information flows from vertex i to vertex j. In this case, vertex i is called the *parent*, vertex j the *child*. And the neighbor set can be expressed as $\mathcal{N}_i = \{j \in \mathcal{V} : (j, i) \in \mathcal{E}, j \neq i\}$.

A *directed path* from vertex i_1 to vertex i_k is a sequence of consecutive directed edges $(i_1, i_2,), (i_2, i_3), \ldots, (i_{k-1}, i_k)$ with distinct vertices. If further $i_1 = i_k$, then it is called a *cycle*. For the special case $i_1 = i_2$, we call it a self-loop. Graph $\mathcal{G}$ is said to be *strongly connected* if for any i and j, there exists a directed path from i to j. It is called *weakly connected* if replacing all of its directed edges with undirected edges produces a connected graph. Any disconnected graph can be partitioned into *weakly connected components* (WCCs). A WCC can be further partitioned into several strongly connected components (SCCs) $(\mathcal{V}_1, \mathcal{E}_1), \ldots, (\mathcal{V}_K, \mathcal{E}_K)$. A SCC is called a *basis SCC* if $\mathcal{G}$ does not have any edges flowing into this SCC from outside. *By substituting each set $\mathcal{V}_k$ with a hyper node v_k^* and drawing an edge from v_k^* to v_j^* if $\mathcal{E}$ contains at least one edge from some vertex in $\mathcal{V}_k$ to some vertex in $\mathcal{V}_j$, we obtain a condensation graph $\mathcal{G}^* = (\mathcal{V}^*, \mathcal{E}^*)$.*

3.2.3 Distributed Consensus-Based Unbiased Estimation Algorithm: Algorithm DCUE

Assuming that proper scheduling is performed such that the nodes update their states synchronously, we propose the following distributed consensus-based unbiased estimation algorithm DCUE. Starting from some deterministic initial value $x_i(0)$, at the end of one round of measurement and transmissions, each SN i updates its estimates $x_i(t)$ as

$$x_i(t+1) = x_i(t) + \rho(t)\alpha_i H_i^T(y_i(t) - H_i x_i(t))$$
$$+ \frac{\rho(t)}{c_i} \left[\sum_{j \in \mathcal{N}_i^S} a_{ij}(x_j(t) - x_i(t)) + \sum_{j \in \mathcal{N}_i^R} a_{ij}(z_j(t) - x_i(t)) \right], \quad (3.3)$$

where $\alpha_i > 0$ is the parameter governing the update rate of information during the estimation process; $\rho(t) > 0$ is the weight; $c_i > 0$ quantifies the confidence of node i's own estimate; $a_{ij} = \sqrt{P_{Tj}|h_{ij}|^2/d_{ij}^n}$ represents the amplitude of the signal received by node i from node j, in which h_{ij} is a fading coefficient describing the channel between nodes i and j^1. Note that, in general, $a_{ij} \neq a_{ji}$, which means that the communication between nodes is asymmetric.

As for RNs, they aggregate their neighboring data as follows

$$z_i(t) = \sum_{j \in \mathcal{N}_i^S} \gamma_{ij} x_j(t) + \sum_{j \in \mathcal{N}_i^R} \gamma_{ij} z_j(t), \ \forall i \in \mathcal{I}_R, \tag{3.4}$$

where $\gamma_{ij} \geq 0$ are nonnegative weights.

In the DCUE algorithm, for a SN, Eq. (3.3) combines the innovation of measurements, $y_i - H_i x_i$, which is the key in Kalman filters [76], and fusion of local estimates from its neighbors, $\sum_{j \in \mathcal{N}_i^S} a_{ij}(x_j - x_i) + \sum_{j \in \mathcal{N}_i^R} a_{ij}(z_j - x_i)$, in the same iteration. This combination is appropriate in applications where the communication rate and the sensing rate of the nodes are similar. While for RNs, the role of them is simply to combine all the incoming data as shown in (3.4), and then rebroadcast them. We emphasize that there are no updates of data at RNs, i.e., the data might be forwarded among the RNs several times before it arrives at one SN, which is then updated to the next-time data.

Remark 3.1 It is clear that the DCUE algorithm is different from the previous algorithms [65, 41, 75, 77]. In these literatures only identical nodes, namely, nodes with respect to SNs in this chapter, are considered. And all the nodes perform exactly the same type of operations, e.g., the $\mathcal{LU}$ algorithm proposed involves only an iteration similar in the form of Eq. (3.3).

To characterize the performance of the DCUE algorithm, we introduce the following two definitions [76].

Definition 3.1 (Asymptotic Unbiasedness) The sequence of estimates $\{x_i(t)\}_{t \geq 0}$ of θ at SN i is said to be asymptotically unbiased if $\lim_{t \to \infty} \mathbb{E}\{x_i(t)\} = \theta$.

Definition 3.2 (Consistency) The sequence of estimates $\{x_i(t)\}_{t \geq 0}$ of θ at SN i is said to be consistent if we have almost surely $\lim_{t \to \infty} x_i(t) = \theta$.

We make the following assumptions throughout the rest of the chapter:

A1 (Connectivity) The communication graph $\mathcal{G}$ is strongly connected.

A2 (Weight rule) For each SN $i \in \mathcal{I}_S$, the weight sequence $\{\rho(t)\}_{t \geq 0}$ satisfies

$$\rho(t) > 0, \ \sum_{t=0}^{\infty} \rho(t) = \infty, \ \sum_{t=0}^{\infty} \rho^2(t) < \infty. \tag{3.5}$$

And for each RN $i \in \mathcal{I}_R$, the weight γ_{ij} satisfies

$$\gamma_{ij} > 0 \Leftrightarrow (j, i) \in \mathcal{E}, \ \sum_{j=1}^{N} \gamma_{ij} = 1. \tag{3.6}$$

---

[1] Some form of channel compensation at the receiver side, e.g., the maximal ratio receiver, is needed to enforce the nonnegativity of a_{ij}, see [75].

A3 (Observability) The observation system (3.1) is distributed observable, i.e., the matrix $\sum_{i=1}^{M} H_i^T H_i$ is of full rank.

Remark 3.2 Assumption **A1** imposes the condition to avoid the trivial case that RNs can not receive any information from SNs. This can be achieved by adopting some deployment strategies of RNs, e.g., [71, 78–80], where RNs are placed on the line segments between SNs to ensure connectivity.

Remark 3.3 Assumption **A3** guarantees that θ is observable from $\{y_1, y_2, \ldots, y_N\}$, although it is not observable at each SN. This condition is essential in distributed estimation problems, which was also used in [41] for homogeneous networks with only SNs.

3.3   Characterization of RNs

One question regarding the relay model (3.4) is that how the relaying feature of RNs can be characterized by (3.4). In this section, we will answer this question.

Note that there are no updates of data at RNs, we only need to consider the data forwarding among nodes in $\check{\mathcal{I}} := \tilde{\mathcal{I}}_S \cup \mathcal{I}_R$, where $\tilde{\mathcal{I}}_S := \{i \in \mathcal{I}_S : \exists\, j \in \mathcal{I}_R, (i, j) \in \mathcal{E} \text{ or } (j, i) \in \mathcal{E}\}$ collects all the SNs that can directly transmit to or receive from any RN. We denote $\tilde{\mathcal{I}}_S$ as the boundary of $\mathcal{I}_S$. The previous question is then related with the amount of information received by any RN or boundary SN that are forwarded from other RNs under (3.4).

3.3.1   Non-cyclic Network

If there are no cycles consisting of RNs, then the RN model (3.4) is similar to the amplify-and-forward relay scheme in [81]. Assume that there is a path from boundary SN j to RN i, say, $j \to k_1 \to k_2 \to \cdots \to k_d \to i$, where $\{k_h, 1 \le h \le d\}$ is the set of RNs, then using (3.4), we can see that along this path the information received by RN i is $\gamma_{j \to i} x_j := \gamma_{ik_d} \cdots \gamma_{k_2 k_1} \gamma_{k_1 j} x_j$. Therefore, at each RN i the received information from all other boundary SNs via RNs can be expressed as

$$\phi_i(t) := \left\{ \sum_{d=1}^{d_l} \sum_{j \to k_1 \to \cdots \to k_d \to i} \gamma_{j \to i} x_j, 1 \le j \le |\tilde{\mathcal{I}}_S| \right\}, \tag{3.7}$$

where d_l is the length of the longest path composed of RNs.

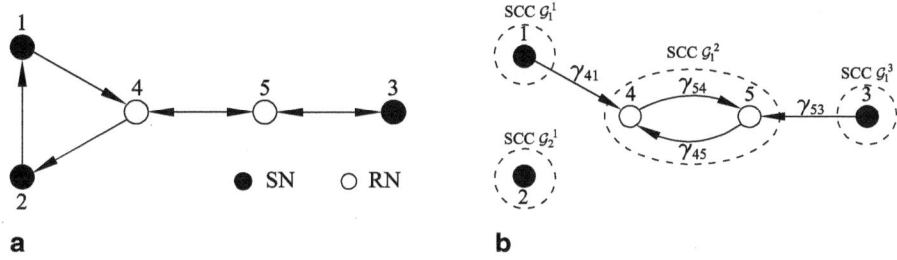

Fig. 3.1 **a** A relay assisted sensor network $\mathcal{G}$. **b** The graph $\mathcal{G}_s[\check{\mathcal{I}}]$

3.3.2 Cyclic Network

In a general relay assisted network, there may exist cycles consisting of RNs. Under model (3.4), the previously forwarded data may be forwarded many times among RNs. In this case, $z_i(t)$ only accounts for temporary information received by each RN i. In order to determine the final data received by RN, a high-dimensional matrix inverse must be solved for such algebra Eq. (3.4). This is generally impractical especially for large-scale networks.

In the following, we will develop a novel graph transformation and then a Markov chain approach to determine the total information forwarded via multi-hop RNs. Construct a directed graph $\mathcal{G}_s[\check{\mathcal{I}}]$ over the set $\check{\mathcal{I}}$ as follows:

- *Induction*: Derive the subgraph induced by $\check{\mathcal{I}}$ via dropping the vertices outside $\check{\mathcal{I}}$ and the associated edges in $\mathcal{G}$. Then delete all the edges $(i, j) \in \tilde{\mathcal{I}}_S \times \tilde{\mathcal{I}}_S$. The resultant graph is denoted by $\mathcal{G}[\check{\mathcal{I}}] = (\check{\mathcal{I}}, \check{\mathcal{E}})$.
- *Weighting*: Associate each edge $(j, i) \in \check{\mathcal{E}}$ with the weight γ_{ij}. Then the matrix $\Gamma = [\gamma_{ij}]$ defines a selfishness graph $\mathcal{G}_s[\check{\mathcal{I}}]$ in the sense that $\gamma_{ij} > 0$ meaning node j always sends its data to node i, where we set $\gamma_{ij} = 0$, $\forall i \in \tilde{\mathcal{I}}_S, j \in \mathcal{I}_R$. Note that even though $\mathcal{G}[\check{\mathcal{I}}]$ is connected, $\mathcal{G}_s[\check{\mathcal{I}}]$ may be disconnected.
- *Partition*: Partition graph $\mathcal{G}_s[\check{\mathcal{I}}]$ into WCCs $\mathcal{G}_k = (\mathcal{V}_k, \mathcal{E}_k)$, $1 \le k \le K$, where K is the number of WCCs. Each WCC $\mathcal{G}_k$ are further decomposed into K_k SCCs $\mathcal{G}_k^j = (\mathcal{V}_k^j, \mathcal{E}_k^j)$, $1 \le j \le K_k$, where K_k is an integer.

Remark 3.4 The graph transformation is meant to partition the whole RNs into separate groups so that we can determine the total information received by any RN that is transmitted from boundary SNs in local regions.

We illustrate the graph transformation using the graph $\mathcal{G}$ shown in Fig. 3.1a. In this example, there are 3 SNs $\mathcal{I}_S = \{1, 2, 3\}$ and 2 RNs $\mathcal{I}_R = \{4, 5\}$. Here the boundary of $\mathcal{I}_S$ is $\tilde{\mathcal{I}}_S = \mathcal{V} = \check{\mathcal{I}}$. The transformed graph $\mathcal{G}_s[\check{\mathcal{I}}]$ is shown in Fig. 3.1b, where two WCCs are present, namely, $\mathcal{G}_1 = \{\mathcal{G}_1^1, \mathcal{G}_1^2, \mathcal{G}_1^3\}$ and $\mathcal{G}_2 = \{\mathcal{G}_2^1\}$. Note that $\mathcal{G}_s[\check{\mathcal{I}}]$ is disconnected, although $\mathcal{G}$ is strongly connected.

Lemma 3.1 *Suppose that* **A1–A2** *hold. Then for each WCC $\mathcal{G}_k$, $\forall 1 \le k \le K$, the total information of each RN in $\mathcal{V}_k$ is determined as a convex combination of the* *estimates of SNs in the same* $\mathcal{V}_k$, *i.e.,*

$$z_{k_i} = conv\{x_i, \ i \in \mathcal{V}_k \cap \tilde{\mathcal{I}}_S\}, \ \forall k_i \in \mathcal{V}_k \backslash \tilde{\mathcal{I}}_S, 1 \le k \le K,$$

where conv$\{\cdot\}$ *is short for the convex combination.*

Proof The main idea of the proof is to group the Laplacian matrix associated with the WCC $\mathcal{G}_k$ into one ordered form according to relations among its SCCs, and then deal with the nodes in each SCC.

For arbitrary WCC $\mathcal{G}_k$, let the number of nodes be m_2, we rearrange the nodes if necessary such that the set of SNs and RNs are labelled by $\mathcal{V}_k^S := \{k_1, k_2, \dots, k_{m_1}\}$ and $\mathcal{V}_k^R := \{k_{m_1+1}, k_{m_1+2}, \dots, k_{m_2}\}$, respectively. It follows from (3.4) and **A2** that for all $k_i \in \mathcal{V}_k^R$,

$$\sum_{j=1}^{m_1} \gamma_{k_i k_j}(z_{k_i} - x_{k_j}) + \sum_{j=m_1+1}^{m_2} \gamma_{k_i k_j}(z_{k_i} - z_{k_j}) = 0. \tag{3.8}$$

Define the Laplacian matrix $L_k = [l_{ij}^k] \in \mathbb{R}^{m_2 \times m_2}$ with entries $l_{ii}^k = \sum_{j=1, j \neq i}^{m_2} \gamma_{k_i k_j}$, and $l_{ij}^k = -\gamma_{k_i k_j}, \ \forall j \neq i$. Since $\mathcal{G}_k$ is a WCC, the edges in its condensation graph $\mathcal{G}_k^*$ can be reordered into the form (v_i^*, v_j^*) with $i \le j$ [75]. Further, from the graph transformation, we find that $\mathcal{G}_k$ has exactly m_1 basis SCCs, namely, $\mathcal{G}_k^1 = (\{k_1\}, \varnothing)$, $\dots, \mathcal{G}_k^{m_1} = (\{k_{m_1}\}, \varnothing)$. This implies that by permutation L_k can be written as a lower block triangular form

$$L_k = \begin{bmatrix} \mathbf{0} & \dots & & \mathbf{0} & \\ B_{(m_1+1)1} & \dots & B_{(m_1+1)(m_1+1)} & & \\ \vdots & & \vdots & \ddots & \\ B_{K_k 1} & \dots & B_{K_k(m_1+1)} & \dots & B_{K_k K_k} \end{bmatrix},$$

where K_k is the number of SCCs of $\mathcal{G}_k$, and for each $1 \le i \le K_k$, $B_{ii} := \tilde{L}_i + D_i$, in which $\tilde{L}_i$ is the Laplacian matrix of the SCC $\mathcal{G}_k^i$, D_i is a nonnegative diagonal matrix with its j-th entry being the sum of weights associated with the incoming edges of node $k_j \in \mathcal{V}_k^i$. Moreover, we have $l_{ii}^k = 1$ for all $m_1 < i \le m_2$ in view of **A2**.

Partition L_k as follows $L_k = [\mathbf{0} \ [E_k \ B_k]^T]^T$, where $E_k = [e_{ij}^k] \in \mathbb{R}^{(m_2-m_1) \times m_1}$ and $B_k = [b_{ij}^k] \in \mathbb{R}^{(m_2-m_1) \times (m_2-m_1)}$. Then, Eq. (3.8) becomes

$$\begin{bmatrix} E_k & B_k \end{bmatrix} \begin{bmatrix} X_k^S \\ Z_k^R \end{bmatrix} = 0, \tag{3.9}$$

where X_k^S and Z_k^R are the stacked column-wise vectors of x_i, $i \in \mathcal{V}_k^S$ and $i \in \mathcal{V}_k^R$, respectively.

In what follows, we will show that B_k is nonsingular. Recall that there are only m_1 basis SCCs of $\mathcal{G}_k$ with the vertex set $\mathcal{V}_k^S$. Let $\mathcal{V}_k^C \subset \mathcal{V}_k^R$ be the set of children of

$\mathcal{V}_k^S$, then it can be seen that the subgraph induced by $\mathcal{V}_k^R$ contains a spanning forest[2] whose roots are given by the set $\mathcal{V}_k^C$. And it is clear that B_k is a diagonally dominant matrix because of the fact $L_k \mathbf{1} = 0$. Moreover, for each $k_j \in \mathcal{V}_k^C$, it is readily shown that $|b_{jj}^k| > \sum_{l \neq j} |b_{jl}^k|$, since $b_{jj}^k = 1$, $\forall 1 \leq j \leq m_2 - m_1$. Thus it follows from Theorem 11 of [82] that B_k is nonsingular. Consequently, we derive from (3.9) that

$$Z_k^R = -B_k^{-1} E_k X_k^S. \tag{3.10}$$

Furthermore, the fact $L_k \mathbf{1} = 0$ gives $-B_k^{-1} E_k \mathbf{1} = \mathbf{1}$.

Now, it only need to verify the nonnegativity of $-B_k^{-1} E_k$. We prove this claim by contradiction. Denote $F_k := -B_k^{-1} E_k = [f_{ij}^k]$. Suppose on the contrary that there exists an entry of F_k such that $f_{j_0 l_0}^k = \min_{j,l} f_{jl}^k < 0$ associated with the node $k_{j_0 + m_1}$ in one SCC $\mathcal{G}_k^{j_0}$. In view of the relation $B_k F_k = -E_k$, we have $\sum_{j \neq j_0} b_{j_0 j}^k f_{j l_0}^k + f_{j_0 l_0}^k = -e_{j_0 l_0}^k$ by noting that $b_{j_0 j_0}^k = 1$. Since $E_k \mathbf{1} + B_k \mathbf{1} = 0$, combining with the structure of L_k yields

$$\sum_{\substack{j=1 \\ j \neq j_0}}^{m_2 - m_1} b_{j_0 j}^k \left(f_{j l_0}^k - f_{j_0 l_0}^k \right) - \sum_{j=1}^{m_1} e_{j_0 j}^k f_{j_0 l_0}^k = -e_{j_0 l_0}^k. \tag{3.11}$$

Note that $b_{j_0 j}^k = l_{(j_0 + m_1)(j + m_1)}^k \leq 0$, $\forall j \neq j_0$ and $\sum_{j=1}^{m_1} e_{j_0 j}^k = \sum_{j=1}^{m_1} l_{(j_0 + m_1)j}^k \leq 0$, hence the left hand side of (3.11) is nonnegative. We have two cases:

Case i) $k_{j_0 + m_1} \in \mathcal{V}_k^C$. By the definition of $\mathcal{V}_k^C$, there is at least one $j_* \in \mathcal{V}_k^S$ such that $l_{(j_0 + m_1)j_*}^k < 0$. Then we have $\sum_{j=1}^{m_1} e_{j_0 j}^k < 0$. Thus the left hand side of (3.11) is negative. But E_k is a nonpositive matrix, which leads to a contradiction.

Case ii) $k_{j_0 + m_1} \notin \mathcal{V}_k^C$. In this case, we have $e_{j_0 j}^k = 0$, $\forall 1 \leq j \leq m_1$, and thus $\sum_{j \neq j_0} b_{j_0 j}^k = 1$. It then follows from (3.11) that there exists $1 \leq j' \leq m_1$ satisfying $f_{j' l_0}^k = f_{j_0 l_0}^k < 0$. Then we can repeat the arguments as above to check whether $k_{j' + m_1} \in \mathcal{V}_k^C$ or not. Since SCC $\mathcal{G}_k^{j_0}$ is strongly connected, under the assumption **A1**, we can always find a directed path $j_0 \to j' \to \cdots \to j^{(s)}$ such that $k_{j^{(s)} + m_1} \in \mathcal{V}_k^C$. For this index $k_{j^{(s)} + m_1}$, a contradiction can be obtained as in case i).

Based on the above analysis, we know that each entry of $-B_k^{-1} E_k$ is nonnegative and all rows sum to 1 since $-B_k^{-1} E_k \mathbf{1} = \mathbf{1}$. It thus follows from (3.10) that each entry of Z_k^R is a convex combination of all entries of X_k^S. This completes the proof. □

Corollary 3.1 *Suppose that **A1–A2** hold. Then for each WCC $\mathcal{G}_k$, the total information of each RN in any hyper-node v_i^* is determines as a convex combination of the states of SNs in $F(i) = \bigcup_{j=0}^{K_k} F^{(j)}(i)$, where $F^{(j)}(i)$ is recursively defined by*

$$F^{(j)}(i) := \begin{cases} \{v_h^* : \exists v_l^* \in F^{(j-1)}(i), (v_h^*, v_l^*) \in \mathcal{E}^*\}, & j \geq 1, \\ \{v_i^*\}, & j = 0. \end{cases}$$

---

[2] A directed tree is a directed graph where all the vertices, except one unique vertex called the root, have exactly one parent. One or more directed trees constitute a directed forest.

Proof From the definition, it is seen that $F(i)$ is the set of all direct and indirect parents of the hyper node v_i^*. Since the condensation graph $\mathcal{G}^* = (\mathcal{V}^*, \mathcal{E}^*)$ has no cycles, it is obvious that the state of $v_{j_1}^* \in F(i)$ is not influenced by the states of all the other hyper nodes $v_{j_2}^* \notin F(i)$.

Repeating the similar arguments as in the proof of Lemma 3.1, we can show that the states of RNs in v_i^* is a convex combination of the states of SNs in $F(i)$ □

By Lemma 3.1, the total information gathered at each RN is a convex combination of those of SNs. Consequently, we can express it as

$$z_i(t) = \sum_{j \in \check{\mathcal{I}}_S} \tilde{\gamma}_{ij} x_j(t), \ \forall i \in \mathcal{I}_R, \tag{3.12}$$

where $0 \le \tilde{\gamma}_{ij} \le 1$ if i and j are both in some WCC $\mathcal{G}_k$, and 0 otherwise, with $\sum_{j \in \check{\mathcal{I}}_S} \tilde{\gamma}_{ij} = 1$.

Using (3.7) and (3.12), we can uniformly express the information gathered at each RN i transmitted from the boundary SNs via RNs as follows

$$\phi_i(t) = \left\{ \sum_{k \in \check{\mathcal{I}}_S} \tilde{\gamma}_{ik} x_k(t) \right\}. \tag{3.13}$$

Remark 3.5 Although (3.13) presents a way to unify the information gathered at each RN and usually $\mathcal{V}_k$ is a much smaller set than $\mathcal{I}_R$, we still need to compute the matrix inverse B_k^{-1} in (3.10) for each WCC $\mathcal{G}_k$ to obtain an explicit expression of $\tilde{\gamma}_{ij}$ for each pair i, j.

In the following, we will provide a graphical interpretation of $\tilde{\gamma}_{ij}$, from which it could be easily determined. The motivation comes from the facts that (1) $\sum_{j \in \check{\mathcal{I}}_S} \tilde{\gamma}_{ij} = 1$ for all $i \in \mathcal{I}_R$ by Lemma 3.1; (2) Every stochastic matrix (where the rows each sum to 1) defines a Markov chain.

3.3.3 Markov Chain Based Characterization

We define the Markov chain $\mathcal{MC}$ as follows:

- *State space*: Let $x_i, i \in \check{\mathcal{I}}_S$ and $z_i, i \in \mathcal{I}_R$ correspond to the states of $\mathcal{MC}$. Rearrange the states orderly such that the state space $\mathcal{S} := \{u_1, u_2, \ldots, u_{|\check{\mathcal{I}}|}\} = \{x_{s_1}, \ldots, x_{s_{|\check{\mathcal{I}}_S|}}, z_{s_{|\check{\mathcal{I}}_S|+1}}, \ldots, z_{s_{|\check{\mathcal{I}}|}}\}$.
- *Transition probability matrix*: In order to define p_{ij} from state u_i to state u_j, we consider two cases (i) $i \le |\check{\mathcal{I}}_S|$ and (ii) $i > |\check{\mathcal{I}}_S|$. For case (i), we set $p_{ii} = 1$, and $p_{ij} = 0$, for all $j \ne i$, while for case (ii), $p_{ii} = 0$ and $p_{ij} = \gamma_{s_i s_j}$, for all $j \ne i$.

In light of **A2**, it easy to see that $\sum_{j=1}^{|\check{\mathcal{I}}|} p_{ij} = 1, \forall 1 \le i \le |\check{\mathcal{I}}|$ and the set of absorbing states is $\mathcal{S}^S := \{u_1, u_2, \ldots, u_{|\check{\mathcal{I}}_S|}\}$. Moreover, by **A1**, we know that $\mathcal{MC}$ defined on the state space $\mathcal{S}$ with transition matrix $P = [p_{ij}]$ is an absorbing Markov chain

[83]. As a result, $\mathcal{S}^R := \{u_{|\check{\mathcal{I}}_S|+1}, u_{|\check{\mathcal{I}}_S|+2}, \ldots, u_{|\check{\mathcal{I}}|}\}$ collects all the transient states of $\mathcal{MC}$.

Theorem 3.1 *Suppose that* **A1–A2** *hold, then* $\{\tilde{\gamma}_{ij}, i \in \mathcal{I}_R, j \in \tilde{\mathcal{I}}_S\}$ *in 3.13 defines a set of probabilities that the process starting in transient states* $\{z_i, i \in \mathcal{I}_R\}$ *ends up in absorbing states* $\{x_j, j \in \tilde{\mathcal{I}}_S\}$ *of the Markov chain* $\mathcal{MC}$.

Proof We emphasize that the transition diagram of $\mathcal{MC}$ constitute the reversal[3] of the graph $\mathcal{G}_s[\check{\mathcal{I}}]$ except the self-loops at absorbing states. Since $\mathcal{G}_s[\check{\mathcal{I}}]$ has K WCCs, we can rewrite the transition matrix P of $\mathcal{MC}$ as $P = \text{diag}\{P_1, P_2, \ldots, P_K\}$, where P_k is the transition matrix corresponding to states $\mathcal{S}_k := \mathcal{S} \cap \mathcal{V}_k$, for all $1 \le k \le K$. Moreover, it follows from (3.9) that P_k can be expressed in the following canonical form

$$
P_k = \begin{bmatrix} I & 0 \\ -E_k & A_k \end{bmatrix}, \forall 1 \le k \le K,
$$

where $-E_k$ and A_k capture transition probabilities from transient states $\mathcal{S}_k^R := \mathcal{S}_k \cap \mathcal{S}^R$ and absorbing states $\mathcal{S}_k^S := \mathcal{S}_k \cap \mathcal{S}^S$ to transient states $\mathcal{S}_k^R$, respectively.

By induction in n, it is straightforward to show that $P^n = \text{diag}\{P_1^n, P_2^n, \ldots, P_K^n\}$, for any integer $n \ge 1$, where

$$
P_k^n = \begin{bmatrix} I & 0 \\ -\sum_{j=0}^{n-1} A_k^j E_k & A_k^n \end{bmatrix}, \forall 1 \le k \le K.
$$

Hence, $-\sum_{j=0}^{n-1} A_k^j E_k$ embodies the n-step transition probabilities from $\mathcal{S}_k^R$ to $\mathcal{S}_k^S$. As an absorbing Markov chain, it follows from Theorem 3.1.1 of [83] that $A_k^n \to 0$ as $n \to \infty$, which implies $\sum_{j=0}^{\infty} A_k^j = (I - A_k)^{-1}$. Therefore, in view of Theorem 3.3.7 of [83], the (i, j)-th entry of $-(I - A_k)^{-1} E_k$ represents the probability that the process starting in some transient state $u_{i'}$ ends up in another absorbing state $u_{j'}$.

On the other hand, within the framework of graph transformation, it follows from the structure of L_k, defined in the proof of Lemma 3.1, that A_k can be expressed as $A_k = I - B_k$ by taking into account the fact that $\sum_{j=1}^{N} \gamma_{ij} = 1$, $\forall i \in \mathcal{I}_R$. By Lemma 3.1, we know that B_k is nonsingular. It thus follows from (3.10) that

$$
Z_k^R = -(I - A_k)^{-1} E_k X_k^S. \tag{3.14}
$$

Comparing (3.14) with (3.12), we find that $\tilde{\gamma}_{ij}$ must be one entry of $-(I - A_k)^{-1} E_k$, that is, $\tilde{\gamma}_{ij}$ characterizes the probability that the process starting in one transient state $u_{i'}$ ends up in the absorbing state $u_{j'}$. This completes the proof.

Remark 3.6 Recall that the transient and absorbing states of $\mathcal{MC}$ corresponds to RNs and boundary SNs in the network, respectively, it is convenient to interpret $\tilde{\gamma}_{ij}$ as

---

[3] The reversal of a directed graph is obtained by reversing the orientation of all the edges.

Fig. 3.2 Sate transition of $\mathcal{MC}$ corresponding to graph $\mathcal{G}$ shown in Fig. 3.1a with u_1, u_2, u_3 being the absorbing states. The numbers beside the curves are the one-step transition probabilities

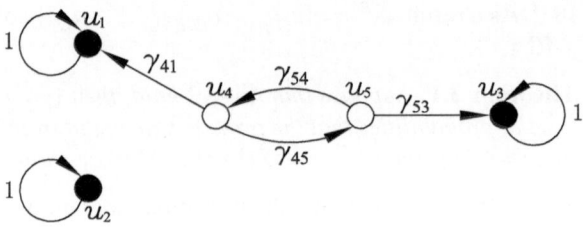

the percentage of information received by RN i which is transmitted from boundary SN j via RNs.

We still use the example shown in Fig. 3.1a for illustration. It can be derived from (3.4) that $\phi_4(t) = \{\tilde{\gamma}_{41}x_1(t) + \tilde{\gamma}_{43}x_3(t)\}$ and $\phi_5(t) = \{\tilde{\gamma}_{51}x_1(t) + \tilde{\gamma}_{53}x_3(t)\}$, where $\tilde{\gamma}_{41} = \gamma_{41}/(1 - \gamma_{45}\gamma_{54})$, $\tilde{\gamma}_{43} = \gamma_{45}\gamma_{53}/(1 - \gamma_{45}\gamma_{54})$, $\tilde{\gamma}_{51} = \gamma_{54}\gamma_{41}/(1 - \gamma_{45}\gamma_{54})$, and $\tilde{\gamma}_{53} = \gamma_{53}/(1 - \gamma_{45}\gamma_{54})$. This indicates that RNs 4 and 5 can successfully receive the percentage of $\tilde{\gamma}_{41}$, $\tilde{\gamma}_{43}$ and $\tilde{\gamma}_{51}$, $\tilde{\gamma}_{53}$ data from SNs 1 and 3, respectively.

As for sensor networks, two intuitive facts are: F.1) the longer the distance between two nodes is, the fewer the information can be successfully adopted, and F.2) the more paths between two nodes, the more information can be received. Our relay model (3.4) embodies these two facts. Consider a special case that $\gamma_{ij} = 1/|\mathcal{N}_i|$, $\forall j \in \mathcal{N}_i$. Note that $\gamma_{41} = \gamma_{45} > \gamma_{45}\gamma_{53}$, we have $\tilde{\gamma}_{41} > \tilde{\gamma}_{43}$. This means that RN 4 receives more information from SN 1 than from SN 3. This is clear from the observation that SN 1 can be reached from RN 4 in one hop, while SN 3 is two hops away. F.2) can be justified as well.

3.3.4   Computing $\tilde{\gamma}_{ij}$ for Steiner-Tree-Based RN Placement

Recently, RN placement in sensor network has attracted a lot of attention, e.g., [70–72, 78, 79], with the main objective of maintaining connectivity of the network. One popular approach is to cast the problem into Steiner minimum tree with minimum number of Steiner points and bounded edge length [71, 78, 79].

Steiner-tree-based RN placement strategies can also be applied to our scenario in this chapter by regarding the SCCs consisting of SNs as hyper nodes. We can then break each edge between hyper nodes into multiple pieces of length at most the communication radius by placing additional RNs, where the distance between two hyper nodes is defined in terms of the distance between two sets. In this way, two basic elements of the relay assisted network are shown in Fig. 3.3. We have the following observation, which can be easily verified for the example shown in Fig. 3.2.

Observation: For each RN i_k, $1 \leq k \leq n$, considering the two basic elements shown in Fig. 3.3,

i) for the former case, it is easy to see that $\tilde{\gamma}_{i_k j} = \gamma_{i_1 j}\gamma_{i_2 i_1} \cdots \gamma_{i_k i_{k-1}}$ and $\tilde{\gamma}_{i_k i} = 0$;

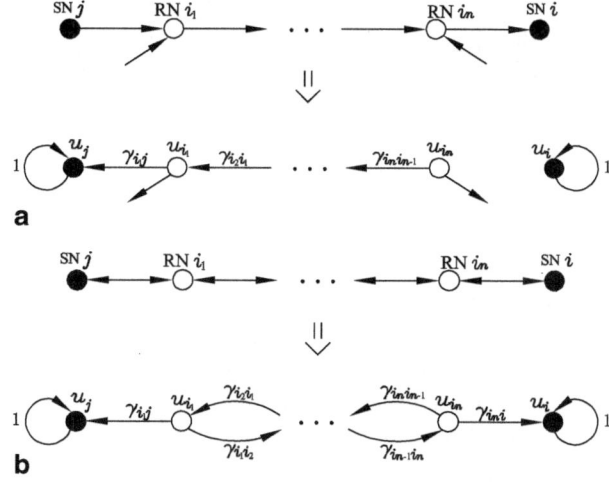

Fig. 3.3 Placement of RNs between two SNs i and j with asymmetric links **a** and symmetric links **b**. The bottom subplots demonstrate the corresponding Markov chains, respectively

ii) for the latter case, we have $\tilde{\gamma}_{i_k j} = \tilde{\gamma}_{i_1 j} \tilde{\gamma}_{i_2 i_1} \cdots \tilde{\gamma}_{i_k i_{k-1}} c_{k,1}$ and the similar form $\tilde{\gamma}_{i_k i} = \tilde{\gamma}_{i_n i} \tilde{\gamma}_{i_{n-1} i_n} \cdots \tilde{\gamma}_{i_k i_{k+1}} c_{k,2}$, where

$$c_{k,1} = \frac{(1 - \gamma_{C_{k+1}}) * \cdots * (1 - \gamma_{C_{n-1}})}{(1 - \gamma_{C_1}) * \cdots * (1 - \gamma_{C_{n-1}})}$$

and

$$c_{k,2} = \frac{(1 - \gamma_{C_1}) * \cdots * (1 - \gamma_{C_{k-2}})}{(1 - \gamma_{C_1}) * \cdots * (1 - \gamma_{C_{n-1}})},$$

and $\gamma_{C_h} = \gamma_{i_h i_{h+1}} \gamma_{i_{h+1} i_h}$, $\forall 1 \le h \le n - 1$. The operator $*$ possesses the same properties with the standard multiplication except that $\gamma_{C_{h_1}} * \gamma_{C_{h_2}} = \gamma_{C_{h_1}} \gamma_{C_{h_2}}$, if $|h_1 - h_2| > 1$, and 0, if $|h_1 - h_2| \le 1$.

Remark 3.7 The above result provides a graphical method to compute $\tilde{\gamma}_{ij}$ at each RN i. The only information required at each RN are the weights of RNs placed on the segment in-between the two SNs connecting them. Actually, the result can be extended to some general cases. We do not elaborate on this topic.

3.4 Convergence Analysis: Asymptotic Unbiasedness and Consistency

Stack the estimates and measurements of all SNs into vectors $x = [x_1^T, x_2^T, \ldots, x_M^T]^T$ and $y = [y_1^T, y_2^T, \ldots, y_M^T]^T$, respectively. For SN $i \in \mathcal{I}_S$, applying (3.13) to (3.3) and recalling that $\sum_{j \in \tilde{\mathcal{I}}_S} \tilde{\gamma}_{ij} = 1, \forall i \in \mathcal{I}_R$, we have the following transformed vector form of (3.3) and (3.4)

$$x(t + 1) = \Psi(t)x(t) + \rho(t)\Upsilon y(t), \qquad (3.15)$$

where $\Psi(t) = I - \rho(t)(\Delta + C^{-1}\hat{L} \otimes I)$, $\Upsilon = (\Phi \otimes I)H^T$, in which $C = \text{diag}\{c_1,\dots,c_M\}$ $\Delta = (\Phi \otimes I)H^T H$, $\Phi = \text{diag}\{\alpha_1,\dots,\alpha_M\}$, $H = \text{diag}\{H_1,\dots,H_M\}$, $\hat{L} = [\hat{l}_{ij}] \in \mathbb{R}^{M \times M}$ with entries

$$
\hat{l}_{ij} = \begin{cases} \sum_{j \in \mathcal{N}_i^S} a_{ij} + \sum_{k \in \mathcal{N}_i^R} a_{ik} \sum_{j \in \tilde{\mathcal{I}}_S \setminus \{i\}} \tilde{\gamma}_{kj}, & j = i, \\[2ex] -a_{ij} - \sum_{k \in \mathcal{N}_i^R} a_{ik}\tilde{\gamma}_{kj}, & j \neq i. \end{cases} \tag{3.16}
$$

Theorem 3.2 *Let $\mathcal{G}$ be a directed graph, then*

(i) $\hat{L}$ is a valid Laplacian matrix;

*(ii) If **A1** is satisfied, then $\hat{L}$ is irreducible, 0 is a simple eigenvalue and its corresponding left eigenvector $\xi = [\xi_1, \xi_2, \dots, \xi_M]^T$ is positive. Moreover, $\Xi\hat{L} + \hat{L}^T\Xi$ is positive semidefinite, where $\Xi = \text{diag}\{\xi_1, \xi_2, \dots, \xi_M\}$.*

Proof (i) By the definition of $\hat{L}$, it can be seen that $\hat{l}_{ii} \geq 0$ and $\hat{l}_{ij} \leq 0$, $\forall i \neq j$. Moreover, we have $\hat{L}\mathbf{1} = 0$. Hence $\hat{L}$ is a valid Laplacian matrix.

(ii) Using Lemma 3.1, by carefully checking different cases, we can show that $\hat{L}$ has exactly one zero eigenvalue if and only if graph $\mathcal{G}$ has a directed tree containing all SNs. The proof is similar to that of Lemma 3.3 of [84], so we omit the details. It also follows that ξ is nonnegative. Moreover, it is obvious that $\Xi\hat{L} + \hat{L}^T\Xi$ has zero row sums. Thus by Gershgorin theorem [85], the positive semidefiniteness is guaranteed.

If **A1** holds, then based on (3.16), we know that $\hat{L}$ is irreducible. Write $\hat{L}$ as $\hat{L} = \max_{1 \leq i \leq M}\hat{l}_{ii}I - \hat{L}'$, then $\hat{L}'^T$ is a nonnegative irreducible matrix and all the columns sum to $\max_{1 \leq i \leq M}\hat{l}_{ii} > 0$, which implies $\rho(\hat{L}'^T) = \max_{1 \leq i \leq M}\hat{l}_{ii}$ [85, Lemma 8.1.21]. Hence $\hat{L}'^T\xi = \rho(\hat{L}'^T)\xi$. By Perron-Frobenius theorem [85], it is guaranteed that $\xi > 0$.

Define the estimation error $e_i(t) := x_i(t) - \theta$ for each SN $i \in \mathcal{I}_S$ and stack $e_i(t)$'s into a vector $e(t) = [e_1^T(t), e_2^T(t), \dots, e_M^T(t)]^T$. Then subtracting $\mathbf{1} \otimes \theta$ from both sides of (3.15) and noticing that $(C^{-1}\hat{L} \otimes I)(\mathbf{1} \otimes \theta) = 0$ by Theorem 3.2, we have

$$
e(t+1) = \Psi(t)e(t) + \rho(t)\Upsilon w(t). \tag{3.17}
$$

In the following, we will establish the asymptotic unbiasedness and consistency of the DCUE algorithm.

3.4.1 Asymptotic Unbiasedness of the DCUE Algorithm

To establish the asymptotic properties, we find that $\hat{L}$ in (3.17) is non-Hermitian in the presence of asymmetric communication. Even if the communication is symmetric, the coexistence of SNs and RNs probably yields a non-Hermitian $\hat{L}$. Hence the approaches for undirected graphs in [73, 74] are not applicable here. Actually, it is hard to analyze the original system (3.17) directly. In this chapter, we propose an appropriate transformation of the original system to circumvent this issue.

By Theorem 3.2, we know that Ξ is positive definite, so ΞC and thus $D :=$ $\sqrt{\Xi C} \otimes I$ are positive definite. With this property, we do the nonsingular transformation $\hat{e}(t) := D\mathbb{E}\{e(t)\}$. Note that $\mathbb{E}\{w(t)\} = 0$, taking expectation on both sides of (3.17) yields the following recursive scheme

$$\hat{e}(t + 1) = D\Psi(t)D^{-1}\hat{e}(t). \tag{3.18}$$

Lemma 3.2 *Suppose that* **A1–A3** *hold, then the matrix* Δ *is positive semidefinite and* $\tilde{\Delta} := \Delta + D(C^{-1}\hat{L} \otimes I)D^{-1} + D^{-1}(\hat{L}^T C^{-1} \otimes I)D$ *is positive definite.*

Proof First note that both Δ and $\tilde{\Delta}$ are symmetric. The positive semidefiniteness of Δ is obvious. Moreover, the properties of Kronecker product implies $D(C^{-1}\hat{L} \otimes I)D^{-1} = (\sqrt{\Xi C})^{-1}\Xi\hat{L}(\sqrt{\Xi C})^{-1} \otimes I$ and $D^{-1}(\hat{L}^T C^{-1} \otimes I)D = (\sqrt{\Xi C})^{-1}\hat{L}^T\Xi(\sqrt{\Xi C})^{-1} \otimes I$. Thus we have

$$\tilde{\Delta} = \Delta + \left((\sqrt{\Xi C})^{-1}(\Xi\hat{L} + \hat{L}^T\Xi)(\sqrt{\Xi C})^{-1}\right) \otimes I$$

$$= \Delta + D^{-1}((\Xi\hat{L} + \hat{L}^T\Xi) \otimes I)D^{-1}$$

By Theorem 3.2, we know that $\Xi\hat{L} + \hat{L}^T\Xi$ and $(\sqrt{\Xi C})^{-1}(\Xi\hat{L} + \hat{L}^T\Xi)(\sqrt{\Xi C})^{-1}$ are both positive semidefinite. As a result, $\tilde{\Delta}$ is positive semidefinite.

In the following, we prove the positive definiteness of $\tilde{\Delta}$ by contradiction. Suppose $\tilde{\Delta}$ is not positive definite, then there exists a nonzero vector $z = [z_1^T, z_2^T, \dots, z_M^T]^T$ $\in \mathbb{R}^{MJ}$ satisfying $z^T\tilde{\Delta}z = 0$, which implies $z^T\Delta z = 0$ and

$$(D^{-1}z)^T((\Xi\hat{L} + \hat{L}^T\Xi) \otimes I)D^{-1}z = 0. \tag{3.19}$$

Since the graph $\mathcal{G}$ is strongly connected, it follows from Theorem 3.2 that $\hat{L}$ is irreducible. Thus Lemma 9 of [86] guarantees that there is a non-zero vector $d \in \mathbb{R}^J$ such that $\frac{1}{\sqrt{\xi_i c_i}}z_i = d, \forall 1 \le i \le M$. Hence

$$0 = z^T\Delta z = \sum_{i=1}^{M}\alpha_i z_i^T H_i^T H_i z_i = \sum_{i=1}^{M}\alpha_i \xi_i c_i d^T H_i^T H_i d. \tag{3.20}$$

Set $\varpi := \min_{1 \le i \le M}\{\alpha_i \xi_i c_i\}$, then $\varpi > 0$ dy Theorem 3.2. Noticing that $H_i^T H_i$ is positive semidefinite for all $1 \le i \le M$, we have

$$\sum_{i=1}^{M}\xi_i c_i d^T \Delta_i d \ge \varpi \sum_{i=1}^{M}d^T H_i^T H_i d \ge 0. \tag{3.21}$$

Equations (3.20) and (3.21) thus give $d^T\sum_{i=1}^{M}H_i^T H_i d = 0$, which contradicts with **A3**. Therefore, $\tilde{\Delta}$ must be positive definite. $\square$

Now we present our first main result about the asymptotic properties of the DCUE algorithm.

Theorem 3.3 *Suppose that* **A1–A3** *hold, then at each SN* $i \in \mathcal{I}_S$, *the estimate sequence* $\{x_i(t)\}_{t \geq 0}$ *is asymptotically unbiased, i.e.,*

$$\lim_{t \to \infty} \mathbb{E}\{x_i(t)\} = \theta, \ \forall i \in \mathcal{I}_S.$$

Proof Continuing the transformed recursion (3.18), $\forall t > t_0$, we have $\hat{e}(t) = \prod_{s=t_0}^{t-1} D\Psi(s)D^{-1}\hat{e}(t_0)$, where $t_0 \geq 0$ is an integer to be determined later. This implies

$$\|\hat{e}(t)\| \leq \prod_{s=t_0}^{t-1} \left\| D\Psi(s)D^{-1} \right\| \|\hat{e}(t_0)\|, \ \forall t > t_0. \tag{3.22}$$

Define the matrices $\Omega := \Delta + C^{-1}\hat{L} \otimes I$, $\hat{\Delta} := \Delta + \tilde{\Delta}$, $\Omega_1 := D^{-1}\Omega^T D^2 \Omega D^{-1}$ and $\Omega_2(t) := \hat{\Delta} - \rho(t)\Omega_1$. Since $D^{-1}\Delta D = D\Delta D^{-1} = \Delta$, it follows that $D^{-1}\Omega^T D + D\Omega D^{-1} = \hat{\Delta}$. Recall that $\Psi(t) = I - \rho(t)\Omega$, the above relation shows

$$D^{-1}\Psi^T(t)D^2\Psi(t)D^{-1} = I - \rho(t)\hat{\Delta} + \rho^2(t)\Omega_1$$
$$= I - \rho(t)\Omega_2(t). \tag{3.23}$$

By Lemma 3.2, $\tilde{\Delta}$ is positive definite, so is $\hat{\Delta}$. It also follows from the definition that Ω_1 is positive semidefinite. Thus $\lambda_{\min}(\hat{\Delta}/2) > 0$ and $\lambda_{\max}(\Omega_1) > 0$. Furthermore, by (3.5), we have $\rho(t) \to 0$ as $t \to \infty$. Therefore, there exists $t_1 \geq 0$ such that $\rho(t)\lambda_{\max}(\Omega_1) \leq \lambda_{\min}(\hat{\Delta}/2)$, whenever $t \geq t_1$. This implies that $\Omega_2(t)$ is positive definite for all $t \geq t_1$. In particular, we derive from the Rayleigh-Ritz theorem [85] that for all $t \geq t_1$,

$$\lambda_{\min}\left(\frac{\hat{\Delta}}{2}\right) \leq \lambda_{\min}(\Omega_2(t)) \leq \lambda_{\max}(\Omega_2(t)) \leq \lambda_{\max}(\hat{\Delta}), \tag{3.24}$$

The fact that $\rho(t) \to 0$ as $t \to \infty$ then guarantees that there exists $t_2 \geq t_1$ so that

$$\rho(t) \leq \frac{1}{\lambda_{\max}(\hat{\Delta})}, \ \forall t \geq t_2. \tag{3.25}$$

The preceding relation along with (3.23) and (3.24) yields

$$\left\| D\Psi(t)D^{-1} \right\|^2 = \left\| D^{-1}\Psi^T(t)D^2\Psi(t)D^{-1} \right\|$$
$$= 1 - \rho(t)\lambda_{\min}(\Omega_2(t)), \ \forall t \geq t_2. \tag{3.26}$$

Put $t_0 = t_2$, then substituting (3.26) to (3.22) gives

$$\|\hat{e}(t)\| \leq \prod_{s=t_0}^{t-1} \sqrt{1 - \rho(t)\lambda_{\min}(\Omega_2(t))}\|\hat{e}(t_0)\|. \tag{3.27}$$

From (3.24) and (3.25), one obtains $0 \leq \rho(t)\lambda_{\min}(\Omega_2(t)) \leq 1$ for all $t > t_0$. The inequality $\ln(1 - \rho(t)\lambda_{\min}(\Omega_2(t))) \leq -\rho(t)\lambda_{\min}(\Omega_2(t))$ thus holds. Hence it follows from (3.24) and (3.27) that for $t > t_0$

$$\|\hat{e}(t)\| \leq \exp\left(-\frac{1}{2}\sum_{s=t_0}^{t-1}\rho(t)\lambda_{\min}(\Omega_2(t))\right)\|\hat{e}(t_0)\|$$

$$\leq \exp\left(-\frac{1}{2}\lambda_{\min}\left(\frac{\hat{\Delta}}{2}\right)\sum_{s=t_0}^{t-1}\rho(t)\right)\|\hat{e}(t_0)\|.$$

Imposing (3.5), one has

$$\lim_{t\to\infty}\left\|\left(\sqrt{\Xi C}\otimes I\right)\mathbb{E}\{e(t)\}\right\| = \lim_{t\to\infty}\|\hat{e}(t)\| = 0. \tag{3.28}$$

Note that $\xi_i c_i > 0$ by Theorem 3.2, Eq. (3.28) then implies that $\|\mathbb{E}\{e(t)\}\|$ converges to zero, the theorem thus follows. From Definition 3.1, we know that the DCUE algorithm produces asymptotic unbiased estimate at each SN. $\quad\square$

3.4.2 Consistency of the DCUE Algorithm

We first give some lemmas which we will find useful in the proof of the consistency. Lemma 3.3 is the well-known Robbins-Siegmund theorem [87]. Lemma 3.4 and Lemma 3.5 provide a way to decompose a quadratic function, which is indispensable to bound the changes of the states of a recursive scheme in the presence of asymmetric communication.

Lemma 3.3 ([87]) *Let $\{\mathcal{F}_t\}_{t\geq 0}$ be a sequence of σ-algebras and $V(t)$, $\mu(t)$, $\nu(t)$ and $\zeta(t)$ be $\mathcal{F}_t$-measurable nonnegative random variables such that for all $t \geq 0$,*

$$\mathbb{E}\{V(t+1)|\mathcal{F}_t\} \leq (1+\mu(t))V(t) + \nu(t) - \zeta(t) \quad a.s.,$$

with $\sum_{t=0}^{\infty}\mu(t) < \infty$ and $\sum_{t=0}^{\infty}\nu(t) < \infty$ almost surely. Then, there exists a nonnegative random variable v^* such that

$$\lim_{t\to\infty}V(t) = v^* \text{ and } \sum_{t=0}^{\infty}\zeta(t) < \infty \quad a.s..$$

Lemma 3.4 *Let $Q(t) := \Psi^T(t)D^2\Psi(t)$. Partition the matrices Δ, $\Psi(t)$ and $Q(t)$ as $\Delta = diag\{\Delta_1,\ldots,\Delta_M\}$, $\Psi(t) = [\Psi_{ij}(t)]$ and $Q(t) = [Q_{ij}(t)]$, respectively, where $\Delta_i = \alpha_i H_i^T H_i$, $\Psi_{ij}(t) \in \mathbb{R}^{J\times J}$ and $Q_{ij}(t) \in \mathbb{R}^{J\times J}$, $\forall 1 \leq i, j \leq M$, then for any* *vector* $v = [v_1^T, v_2^T, \ldots, v_M^T]^T \in \mathbb{R}^{MJ}$, *we have*

$$v^T Q(t) v = \sum_{i=1}^{M} v_i^T \left(\rho^2(t) \sum_{k=1}^{M} \xi_k \hat{l}_{ki} \Delta_k + \xi_i c_i \, (I - \rho(t)\Delta_i)^2 \right) v_i$$

$$- \frac{1}{2} \sum_{i=1}^{M} \sum_{\substack{j=1 \\ j \neq i}}^{M} (v_j - v_i)^T Q_{ij}(t)(v_j - v_i).$$

Proof By the definition of $Q(t)$, we have

$$\sum_{\substack{j=1 \\ j \neq i}}^{M} Q_{ij}(t) = \sum_{\substack{k=1 \\ k \neq i}}^{M} \xi_k c_k \Psi_{ki}^T(t) \left(\Psi_{kk}(t) + \sum_{\substack{j=1 \\ j \neq i,k}}^{M} \Psi_{kj}(t) \right) + \xi_i c_i \Psi_{ii}^T(t) \sum_{\substack{j=1 \\ j \neq i}}^{M} \Psi_{ij}(t).$$

Further, $\Psi_{ii}(t) = I - \rho(t)(\Delta_i + \hat{l}_{ii} I / c_i)$ and $\Psi_{ij}(t) = -\rho(t)\hat{l}_{ij} I / c_i, \, \forall j \neq i$. For each $1 \leq i \leq M$, let $U_i := \Delta_i + \hat{l}_{ii} I / c_i$, then we have

$$\sum_{\substack{j=1 \\ j \neq i}}^{M} Q_{ij}(t) = -\rho(t) \sum_{\substack{k=1 \\ k \neq i}}^{M} \xi_k \hat{l}_{ki} \left(I - \rho(t) U_k - \frac{\rho(t)}{c_k} \sum_{\substack{j=1 \\ j \neq i,k}}^{M} \hat{l}_{kj} I \right)$$

$$- \rho(t) \xi_i \, (I - \rho(t) U_i) \sum_{\substack{j=1 \\ j \neq i}}^{M} \hat{l}_{ij}. \tag{3.29}$$

Recall that $\hat{L}$ is a Laplacian matrix, Eq. (3.29) can be rewritten as

$$\sum_{\substack{j=1 \\ j \neq i}}^{M} Q_{ij}(t) = -\rho(t) \sum_{\substack{k=1 \\ k \neq i}}^{M} \xi_k \hat{l}_{ki}(I - \rho(t)\Delta_k) - \rho^2(t) \sum_{\substack{k=1 \\ k \neq i}}^{M} \frac{\xi_k}{c_k} \hat{l}_{ki}^2 I$$

$$+ \rho(t) \xi_i \hat{l}_{ii} \, (I - \rho(t) U_i).$$

On the other hand, we have for all $1 \leq i \leq M$

$$Q_{ii}(t) = \sum_{k=1}^{M} \xi_k c_k \Psi_{ki}^T(t) \Psi_{ki}(t)$$

$$= \rho^2(t) \sum_{\substack{k=1 \\ k \neq i}}^{M} \frac{\xi_k}{c_k} \hat{l}_{ki}^2 I + \xi_i c_i \, (I - \rho(t) U_i)^2.$$

Thus the preceding two relations yield

$$Q_{ii}(t) = -\sum_{\substack{j=1 \\ j \neq i}}^{M} Q_{ij}(t) - \rho(t) \sum_{\substack{k=1 \\ k \neq i}}^{M} \xi_k \hat{l}_{ki}(I - \rho(t)\Delta_k) + \xi_i c_i (I - \rho(t) U_i)(I - \rho(t)\Delta_i)$$

$$= -\sum_{\substack{j=1 \\ j \neq i}}^{M} Q_{ij}(t) + \rho^2(t) \sum_{k=1}^{M} \xi_k \hat{l}_{ki} \Delta_k + \xi_i c_i (I - \rho(t)\Delta_i)^2,$$

where in the last step the fact that $\xi^T \hat{L} = 0$ was used.

It thus follows that for arbitrary vector $v = [v_1^T, v_2^T, \ldots, v_M^T]^T \in \mathbb{R}^{MJ}$,

$$
\begin{aligned}
v^T Q(t) v &= \sum_{i=1}^{M} v_i^T \sum_{\substack{j=1 \\ j \neq i}}^{M} Q_{ij}(t) v_j + \sum_{i=1}^{M} v_i^T Q_{ii}(t) v_i \\
&= \sum_{i=1}^{M} v_i^T \sum_{\substack{j=1 \\ j \neq i}}^{M} Q_{ij}(t)(v_j - v_i) + \rho^2(t) \sum_{i=1}^{M} v_i^T \sum_{k=1}^{M} \xi_k \hat{l}_{ki} \Delta_k v_i \\
&\quad + \sum_{i=1}^{M} \xi_i c_i v_i^T (I - \rho(t)\Delta_i)^2 v_i.
\end{aligned}
\tag{3.30}
$$

By the symmetry property of $Q(t)$, it can be verified that

$$\sum_{i=1}^{M} v_i^T \sum_{\substack{j=1 \\ j \neq i}}^{M} Q_{ij}(t)(v_j - v_i) = -\frac{1}{2} \sum_{\substack{j=1 \\ j \neq i}}^{M} (v_j - v_i)^T Q_{ij}(t)(v_j - v_i).$$

Inserting the previous identity into (3.30) finally completes the proof. □

Lemma 3.5 *For any $j \neq i$, denote $\tilde{Q}_{ij}(t) := Q_{ij}(t)/\rho(t)$, then $\lim_{t\to\infty} \tilde{Q}_{ij}(t) = -(\xi_i \hat{l}_{ij} + \xi_j \hat{l}_{ji})I$. If further A1 holds, then there exists $t^* > 0$ such that $\tilde{Q}_{ij}(t)$ is positive semidefinite, whenever $t \geq t^*$.*

Proof It follows from the definition of $Q(t)$ that for any $j \neq i$,

$$\tilde{Q}_{ij}(t) = \rho(t)\left(\sum_{k=1}^{M} \frac{\xi_k}{c_k} \hat{l}_{ki} \hat{l}_{kj} I + \xi_i \hat{l}_{ij} \Delta_i + \xi_j \hat{l}_{ji} \Delta_j \right) - (\xi_i \hat{l}_{ij} + \xi_j \hat{l}_{ji})I.$$

It is clear that $\tilde{Q}_{ij}(t)$ is symmetric and $\lim_{t\to\infty} \tilde{Q}_{ij}(t) = -(\xi_i \hat{l}_{ij} + \xi_j \hat{l}_{ji})I$. For any $j \neq i$, consider the next two cases:

Case i) $\hat{l}_{ij} \neq 0$ or $\hat{l}_{ji} \neq 0$. In this case, one has $\hat{l}_{ij} < 0$ or $\hat{l}_{ji} < 0$. By Theorem 3.2, we thus have $\xi_i \hat{l}_{ij} + \xi_j \hat{l}_{ji} < 0$. By A2, it can shown that $\rho(t) \to 0$ as $t \to \infty$. As a result, there exists $t^* > 0$ so that $\tilde{Q}_{ij}(t)$ is positive definite, for all $t \geq t^* > 0$.

Case ii) $\hat{l}_{ij} = \hat{l}_{ji} = 0$. In this case, we have $\tilde{Q}_{ij}(t) = \rho(t) \sum_{k \neq i,j} \xi_k \hat{l}_{ki} \hat{l}_{kj} I / c_k$. Recalling that $\hat{l}_{ki} \hat{l}_{kj} \geq 0$ and $\xi_k > 0$ by Theorem 3.2, $\forall k \neq i, j$. This implies that $\tilde{Q}_{ij}(t)$ is positive semidefinite for all $t \geq 0$.

Combining the above two cases completes the proof. □

Based on the above lemmas, the consistency property of the DCUE algorithm is given by the following theorem.

Theorem 3.4 *Suppose that* **A1–A3** *hold, then the estimate sequence* $\{x_i(t)\}_{t\geq 0}$ *of each SN* $i \in \mathcal{I}_S$ *satisfies*

$$\lim_{t\to\infty} x_i(t) = \theta \quad a.s.$$

Proof Define the σ-algebra $\mathcal{F}_t := \sigma\{w(s), s < t\}$, then $w(t)$ is independent of $\mathcal{F}_t$. Thus $\mathbb{E}\{w(t)|\mathcal{F}_t\} = \mathbb{E}\{w(t)\} = 0$.

Consider the stochastic process $V(t) := \sum_{i=1}^{M} \xi_i c_i \|e_i\|^2$. It can be expressed in the vector form $V(t) = e^T(t)(\varXi C \otimes I)e(t)$ in terms of Kronecker product. Use (3.17) to write

$$V(t+1) = e^T(t)Q(t)e(t) + 2\rho(t)e^T(t)\Psi^T(t)D^2 \varUpsilon w(t) + \rho^2(t)w^T(t)\varUpsilon^T D^2 \varUpsilon w(t).$$

Taking expectation conditioned on $\mathcal{F}_t$ on both sides of the above equation, we obtain

$$\mathbb{E}\{V(t+1)|\mathcal{F}_t\} = e^T(t)Q(t)e(t) + \rho^2(t)\mathrm{tr}\left(\varUpsilon^T D^2 \varUpsilon W\right).$$

where use was made of the relation $w^T(t)\varUpsilon^T D^2 \varUpsilon w(t) = \mathrm{tr}\left(\varUpsilon^T D^2 \varUpsilon w(t)w^T(t)\right)$. It follows from Lemma 3.4 that

$$\mathbb{E}\{V(t+1)|\mathcal{F}_t\} = V(t) + \rho^2(t)\sum_{i=1}^{M} e_i^T(t)F_i e_i(t) - \rho(t)V^o(t)$$

$$+ \rho^2(t)\mathrm{tr}\left(\varUpsilon^T D^2 \varUpsilon W\right), \tag{3.31}$$

where $F_i = \xi_i c_i \Delta_i^2 + \sum_{k=1}^{M} \xi_k \hat{l}_{ki} \Delta_k$, $V^o(t) = 1/2V^{oo}(t) + 2\sum_{i=1}^{M} \xi_i c_i e_i^T(t)\Delta_i e_i(t)$ and $V^{oo}(t) = \sum_{i=1}^{M} \sum_{j\neq i} (e_j(t) - e_i(t))^T \widetilde{Q}_{ij}(t)(t)(e_j(t) - e_i(t))$.

We will show that for sufficiently large $t > 0$, $V(t)$ falls under the purview of Lemma 3.3. For this purpose, we next estimate $\sum_{i=1}^{M} e_i^T(t)F_i e_i(t)$ and $V^o(t)$.

(i) *Estimate of* $\sum_{i=1}^{M} e_i^T(t)F_i e_i(t)$: By Rayleigh-Ritz theorem [85], we find that

$$\lambda_{\max}(F_i) = \max_{z\neq 0} \frac{z^T F_i z}{z^T z} \geq \max_{z\neq 0} \frac{\sum_{k=1}^{M} \xi_k \hat{l}_{ki} z^T \Delta_k z}{z^T z},$$

since Δ_i is positive semidefinite for all $1 \leq i \leq M$. Fix a nonzero vector $z_* \in \mathbb{R}^J$, recalling that $\hat{l}_{ki} \leq 0, \forall k \neq i$ and $\xi^T \hat{L} = 0$, one has

$$\sum_{k=1}^{M} \xi_k \hat{l}_{ki} z_*^T \Delta_k z_* \geq \xi_i \hat{l}_{ii} \left(z_*^T \Delta_i z_* - \max_{\substack{1\leq k\leq M \\ k\neq i}} z_*^T \Delta_k z_*\right).$$

Pick $i_0 = \arg\max_{1\leq i\leq M} z_*^T \Delta_i z_*$. Since $\xi_{i_0} \hat{l}_{i_0 i_0} > 0$, the previous inequalities imply

$$\max_{1\leq i\leq M} \lambda_{\max}(F_i) \geq \max_{\substack{1\leq i\leq M \\ z\neq 0}} \frac{\sum_{k=1}^{M} \xi_k \hat{l}_{ki} z^T \Delta_k z}{z^T z}$$

$$\geq \frac{\sum_{k=1}^{M} \xi_k \hat{l}_{ki0} z_*^T \varDelta_k z_*}{z_*^T z_*} \geq 0.$$

Denote $\chi := \frac{\max_{1 \leq i \leq M} \lambda_{\max}(F_i)}{\min_{1 \leq i \leq M} \xi_i c_i}$, then $\chi \geq 0$ and thus

$$\sum_{i=1}^{M} e_i^T(t) F_i e_i(t) \leq \chi V(t). \tag{3.32}$$

(ii) *Estimate of* $V^o(t)$: In view of Lemma 3.5, we know that $V^{oo}(t) \geq 0$ and $\min_{Q_{ij} \neq 0} \{\lambda_{\min}(\tilde{Q}_{ij}(t))\} > 0$, $\forall t \geq t^*$. Recalling that $\varDelta_i$ is positive semidefinite and $\xi_i > 0$, $\forall 1 \leq i \leq M$ in light of Theorem 3.2, it follows that $V^o(t) \geq 0$, $\forall t \geq t^*$. Moreover, for any $e(t) \neq 0$, we have $V^o(t) > 0$, $\forall t \geq t^*$. To see this, it suffices to show that we cannot have simultaneously

$$\sum_{i=0}^{M} \xi_i c_i e_i^T(t) \varDelta_i e_i(t) = 0 \text{ and } V^{oo}(t) = 0.$$

Indeed if $V^{oo}(t) = 0$, then in view of $V^{oo}(t) \geq \min_{Q_{ij} \neq 0} \{\lambda_{\min}(\tilde{Q}_{ij}(t))\} \sum_{i=0}^{M} \sum_{j \neq i} \|e_j(t) - e_i(t)\|^2$, we must have $e_i(t) = e_j(t)$, $\forall i \neq j$, since $\mathcal{G}$ is strongly connected by **A1**. So if we also have $\sum_{i=0}^{M} \xi_i c_i e_i^T(t) \varDelta_i e_i(t) = 0$, then $\sum_{i=0}^{M} \xi_i c_i \varDelta_i = \sum_{i=0}^{M} \xi_i c_i \alpha_i H_i^T H_i$ is singular. Similar to (3.21), we can derive a contradiction with **A3**.

Denote $\mu(t) := \rho^2(t)\chi$, $\nu(t) := \rho^2(t)\text{tr}(\Upsilon^T D^2 \Upsilon W)$ and $\zeta(t) := \rho(t)V^o(t)$. Obviously, $\mu(t)$, $\nu(t)$ and $\zeta(t)$ are all nonnegative for all $t \geq t^*$. And it is clear that $\sum_{t=0}^{\infty} \mu(t) < \infty$ and $\sum_{t=0}^{\infty} \nu(t) < \infty$ by (3.5). Hence applying Lemma 3.3 to $V(t)$ gives that there is a nonnegative random variable $v^* \geq 0$ so that $\lim_{t \to \infty} V(t) = v^*$ a.s. and

$$\sum_{t=t^*}^{\infty} \zeta(t) = \sum_{t=t^*}^{\infty} \rho(t)V^o(t) < \infty \text{ a.s.}. \tag{3.33}$$

We claim that there exists a subsequence $\{t_1, t_2, \dots\} \subset \{1, 2, \dots\}$ such that $\mathbb{P}\{\lim_{k \to \infty} V(t_k) = 0\} = 1$. In fact, assuming otherwise that there is a constant $\underline{v} > 0$ so that $\mathbb{P}\{\mathcal{D}\} \geq \underline{v}$, where $\mathcal{D} = \{\omega : \liminf_{k \to \infty} V(t_k, \omega) \geq 2\underline{v}\}$. For any event $\omega \in \mathcal{D}$, we can find a random variable $t(\underline{v}, \omega) \geq t^*$ such that $V(t, \omega) \geq \underline{v}$, whenever $t \geq t(\underline{v}, \omega)$. Since $V^o(t) > 0$ for any $e(t) \neq 0$, it follows that there exists a constant $\underline{v}^o > 0$ so that $V^o(t, \omega) \geq \underline{v}^o$, $\forall t \geq t(\underline{v}, \omega)$. This fact gives

$$\sum_{t=t^*}^{\infty} \rho(t)V^o(t, \omega) \geq \underline{v}^o \sum_{t=t(\underline{v},\omega)}^{\infty} \rho(t) = \infty,$$

which is in contradiction to (3.33).

The above contradiction implies that $V(t) \to 0$ a.s. as $t \to \infty$. By Theorem 3.2, the matrix ΞC is positive definite. Therefore, $\mathbb{P}\{\lim_{t \to \infty} e(t) = 0\} = 1$. In terms of Definition 3.2, the consistency of the DCUE algorithm is established. $\square$

3.5 Rate of Convergence

Besides the convergence, we are particularly interested in the rate of convergence of the DCUE algorithm. We adopt the sum of squared errors from the true value θ, i.e., $\sum_{i=1}^{M} \|x_i(t) - \theta\|^2$, as an indicator to measure how far we are from the convergence.

3.5.1 Rate of Convergence of the DCUE Algorithm

The next theorem provides a quantitative bound on the rate of convergence of $\sum_{i=1}^{M} \|x_i(t) - \theta\|^2$.

Theorem 3.5 *Suppose that* **A1–A3** *hold, further,* $\rho(t)$ *has the following form*

$$\rho(t) = \frac{b}{(t+1)^\beta}, \quad \frac{1}{2} < \beta \leq 1, \; b > 0, \; \forall t \geq 0,$$

then for any $0 \leq \kappa < \min\{2\beta - 1, b\delta\}$, *where* $\delta = \frac{\lambda_{\min}((\varXi \hat{L} + \hat{L}^T \varXi) \otimes I + 2D^2 \Delta)}{2\max_{1 \leq i \leq M} \xi_i c_i}$, *we have for all large* $t \geq t^*$

$$\sum_{i=1}^{M} \|x_i(t) - \theta\|^2 = o(t^{-\kappa}) \; a.s.$$

Proof We first show that $V^o(t) \geq \delta V(t), \forall t \geq t^*$. To this end, define a Laplacian-like matrix $L^o(t) = [L_{ij}^o(t)]$, where $L_{ij}^o(t) \in \mathbb{R}^{J \times J}, \forall 1 \leq i, j \leq M$ as follows

$$L_{ij}^o(t) = \begin{cases} \sum_{j=1, j \neq i}^{M} \tilde{Q}_{ij}(t), & j = i, \\ -\tilde{Q}_{ij}(t), & j \neq i. \end{cases}$$

It follows from Lemma 3.5 that for any $j \neq i$, $L_{ij}^o(t) \to (\xi_i \hat{l}_{ij} + \xi_j \hat{l}_{ji})I$ as $t \to \infty$. After some calculation it shows that $L^o(t) \to (\varXi \hat{L} + \hat{L}^T \varXi) \otimes I$ as $t \to \infty$.

Note that $\tilde{Q}_{ij}(t) = \tilde{Q}_{ji}(t)$, $\forall j \neq i$, it can be verified that $V^{oo}(t) = 2e^T(t)L^o(t)e(t)$. As a result, we can obtain $V^o(t) = e^T(t)(L^o(t) + 2D^2\Delta)e(t)$. Therefore,

$$V^o(t) - \delta V(t) \geq \left(\lambda_{\min}(L^o(t) + 2D^2\Delta) - \delta \max_{1 \leq i \leq M} \xi_i c_i\right)\|e(t)\|^2.$$

Based on the previous analysis, we find that $L^o(t) + 2D^2\Delta \to L_\infty$ as $t \to \infty$, where the limit $L_\infty = (\varXi \hat{L} + \hat{L}^T \varXi) \otimes I + 2D^2\Delta$. It follows from Lemma 3.2 that $L_\infty = D(\tilde{\Delta} + \Delta)D$ is positive definite. Since the eigenvalues of a matrix depend continuously upon its entries [85, Appendix D], it follows that for sufficiently large $t \geq t^*$, $\lambda_{\min}(L^o(t) + 2D^2\Delta) \geq 1/2\lambda_{\min}(D(\tilde{\Delta} + \Delta)D)$. And thus the claim $V^o(t) \geq \delta V(t), \forall t \geq t^*$ follows.

By Theorem 3.4, the boundedness of $V(t)$ is guaranteed, say, $\mathbb{P}\{\sup_{t \geq 0} V(t) \leq \bar{v}\} = 1$, for some constant $\bar{v} > 0$. This together with (3.31) and (3.32) gives for all $t \geq t^*$,

$$\mathbb{E}\{V(t+1)|\mathcal{F}_t\} \leq \left(1 - \frac{b\delta}{(t+1)^\beta}\right) V(t) + \frac{\tilde{b}}{(t+1)^{2\beta}}, \tag{3.34}$$

where $\tilde{b} = b^2(\chi\bar{v} + \text{tr}(\Upsilon^T D^2 \Upsilon W))$.

Let $\tilde{V}(t) := t^\kappa V(t)$ and $g(t) := 1 - \left(1 + \frac{1}{t}\right)^\kappa \left(1 - \frac{b\delta}{(t+1)^\beta}\right)$, then it follows from (3.34) that

$$\mathbb{E}\{\tilde{V}(t+1)|\mathcal{F}_t\} \leq \tilde{V}(t) - g(t)\tilde{V}(t) + \frac{\tilde{b}}{(t+1)^{2\beta-\kappa}}, \quad \forall t \geq t^*.$$

Write $g(t)$ as

$$g(t) = \frac{1}{(t+1)^\beta}\left[b\delta\left(1+\frac{1}{t}\right)^\kappa - (t+1)^\beta\left(\left(1+\frac{1}{t}\right)^\kappa - 1\right)\right].$$

It can be verified that

$$\lim_{t \to \infty} (t+1)^\beta\left(\left(1+\frac{1}{t}\right)^\kappa - 1\right) = \begin{cases} \kappa, & \beta = 1, \\ 0, & \beta < 1. \end{cases}$$

Consequently, for large enough $t > t^*$, we have $g(t) \geq 2^{-1}(b\delta - \kappa)(t+1)^{-\beta} > 0$ and clearly $\sum_{t=t^*}^{\infty} g(t) = \infty$. On the other hand, noticing that $2\beta - \kappa > 1$, we have $\sum_{t=t^*}^{\infty} (t+1)^{\kappa-2\beta} < \infty$. Lemma 3.3 thus guarantees that $\lim_{t \to \infty} \tilde{V}(t)$ exists almost surely and $\sum_{t=t^*}^{\infty} g(t)\tilde{V}(t) < \infty$.

Following the similar line as in the proof of Theorem 3.4, we can show that $\tilde{V}(t) \to 0$ a.s. as $t \to \infty$. Note that $V(t) \geq \min_{1 \leq i \leq M} \xi_i c_i \sum_{i=1}^{M} \|x_i(t) - \theta\|^2$ and $\xi_i c_i > 0$ for all $1 \leq i \leq M$ by Theorem 3.2. Therefore

$$\lim_{t \to \infty} t^\kappa \sum_{i=1}^{M} \|x_i(t) - \theta\|^2 = 0 \text{ a.s.},$$

from which the theorem follows.   □

Remark 3.8 We can choose the parameter b so that $b > \frac{2\max_{1 \leq i \leq M} \xi_i c_i}{\lambda_{\min}((\Xi\hat{L}+\hat{L}^T\Xi)\otimes I + 2D^2\Delta)}$ and $\beta = 1$. In this case, $0 \leq \kappa < 1$, and the convergence rate of $\|x(t) - \theta\mathbf{1}\|$ can be achieved at $o(t^{-\kappa'})$, for arbitrary $\kappa' \in [0, 1/2)$, since $\kappa' = \kappa/2$.

3.5.2   Rate of Convergence of Corresponding Consensus Algorithms

In this subsection, we will compare the convergence rate of the consensus algorithm derived from the DCUE algorithm with that of the standard consensus algorithms.

For the homogeneous networks with only SNs, the standard scalar consensus algorithms in the literature have the following basic form [88]

$$x_i(t+1) = x_i(t) + \varepsilon \sum_{j \in \mathcal{N}_i} a_{ij}(x_j(t) - x_i(t)), \ i \in \mathcal{V}, \tag{3.35}$$

where $0 < \varepsilon < 1/(\max_i \sum_{j \in \mathcal{N}_i} a_{ij})$. Use the notation $\tilde{x} = [x^T, x_{M+1}, \ldots, x_N]^T \in \mathbb{R}^N$ to write (3.35) in the compact form, we obtain

$$\tilde{x}(t+1) = (I - \varepsilon L)\tilde{x}(t),$$

where L and $\mathcal{A} = [a_{ij}]$ are the Laplacian matrix and adjacent matrix corresponding to the graph over all N SNs, respectively. On the other hand, the consensus algorithm associated with the DCUE algorithm (3.3) and (3.4) is given by

$$x(t+1) = (I - \varepsilon \hat{L})x(t).$$

Assume that the communications are symmetric, that is, the corresponding graphs are undircted, then the convergence rates [88] of the two algorithms are defined by

$$\text{rt}(L) := \max_{z \perp \mathbf{1}, \|z\|=1} z^T (I - \varepsilon L)z$$

and

$$\text{rt}(\hat{L}) := \max_{z \perp \mathbf{1}, \|z\|=1} z^T (I - \varepsilon \hat{L})z,$$

respectively. The smaller the value of the rate of convergence, the better the performance of the algorithm.

Define the matrix $\Lambda = [g_{ij}] \in \mathbb{R}^{M \times M}$ with entries $g_{ii} = -\sum_{k \in \mathcal{N}_i^R} a_{ik}\tilde{\gamma}_{ki}$ and $g_{ij} = \sum_{k \in \mathcal{N}_i^R} a_{ik}\tilde{\gamma}_{kj}, \forall j \neq i$. We obtain the next theorem.

Theorem 3.6 *The gap between the convergence rates $\text{rt}(\hat{L})$ and $\text{rt}(L)$ is $\text{rt}(\hat{L}) - \text{rt}(L) = \varepsilon(\lambda_2(L) - \lambda_2(\hat{L}))$, and*

$$-\varepsilon(\lambda_{\max}(\Lambda) + \lambda_2(L_M) - \lambda_2(L)) \leq \text{rt}(\hat{L}) - \text{rt}(L) \leq -\varepsilon\lambda_{\min}(\Lambda),$$

where L_M is the principal submatrix by deleting the last $N - M$ rows and columns from L.

Proof First note that $L\mathbf{1} = \hat{L}\mathbf{1} = 0$, then by definition, one obtains $\text{rt}(L) = 1 - \varepsilon\lambda_2(L)$ and $\text{rt}(\hat{L}) = 1 - \varepsilon\lambda_2(\hat{L})$. This shows that $\text{rt}(L)$ is $\text{rt}(\hat{L}) - \text{rt}(L) = \varepsilon(\lambda_2(L) - \lambda_2(\hat{L}))$.

From the definition of $\hat{L}$ in (3.16) and noting that $\sum_{j \in \tilde{\mathcal{I}}_S} \tilde{\gamma}_{ij} = 1, \forall i \in \mathcal{I}_R$, we can express $\hat{L}$ as $\hat{L} = L_M + \Lambda$, where L_M is the principal submatrix by deleting the last $N - M$ rows and columns from L. Both L_M and Λ are symmetric, it then follows from Weyl's theorem [85] that $\lambda_2(L_M) + \lambda_{\min}(\Lambda) \leq \lambda_2(\hat{L}) \leq \lambda_2(L_M) + \lambda_{\max}(\Lambda)$. On

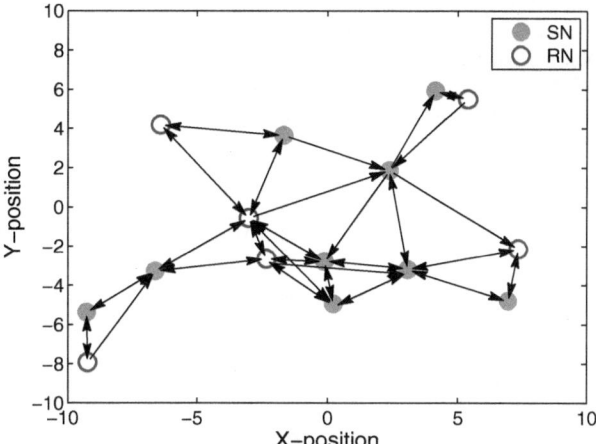

Fig. 3.4 An asymmetric communication graph with nine SNs and six RNs

the other hand, as a principal submatrix of L, we have $\lambda_2(L) \le \lambda_2(L_M)$ [85, Theorem 4.3.15]. Combining the above two inequalities then completes the proof. □

Remark 3.9 The placement of RNs do contribute to the convergence rate, provided that some appropriate replacement strategy of SNs with RNs is adopted such that $\lambda_2(L) \le \lambda_2(\hat{L})$.

3.6 Numerical Studies

In this section, numerical studies are presented to validate the performance of the DCUE algorithm.

We consider a fixed relay assisted sensor network composed of nine SNs $\{1, 2, \ldots, 9\}$ and six RNs $\{10, 11, \ldots, 15\}$, shown in Fig. 3.4, to cooperatively estimate the unknown parameter $\theta = [1, 5]^T$. The network is obtained by distributing all the nodes randomly in the square $[-10, 10] \times [-10, 10]$ with inter-sensor distances no smaller than two to meet the coverage requirement. To simplify the simulation, we adopt a simple model for path loss without considering path fading $a_{ij} = \sqrt{P_{Tj}/d_{ij}^\eta}$, where P_{Tj} is randomly chosen from the interval $[10, 20]$. The node j can communicate with node i if $a_{ij} > 0.7$. Weights for RNs are set to be $\gamma_{ij} = a_{ij} / \sum_{j \in \mathcal{N}_i} a_{ij}$, $\forall 10 \le i \le 15$.

The measurements at SNs are $y_i(t) = H_i\theta + w_i(t)$, where $H_1 = H_2 = H_5 = H_6 = H_9 = [1, 0]$, $H_3 = H_4 = H_7 = H_8 = [0, 1]$, and $w_i(t)$ is the white observation noise with variance $\sigma_i^2 = 20$, for all $1 \le i \le 9$. Clearly, θ is not observable by individual SN, but is distributed observable by all SNs, since $\sum_{i=1}^9 H_i^T H_i = \text{diag}\{5, 4\}$ is positive definite.

Algorithm DCUE is implemented for the above setup with $\alpha_i = 5$, $c_i = 0.5$, $\rho(t) = 0.6/(t+1)$, and the initial conditions $x_i(0) = [x_{i1}(0), x_{i2}(0)]^T$, where $x_{i1}(0)$

Fig. 3.5 a Estimates of
$\theta = [1, 5]^T$ of all the SNs. **b**
Normalized estimation error
of SNs with $\min_i |\mathcal{N}_i|$,
$\max_i |\mathcal{N}_i|$ and centralized
BLUE

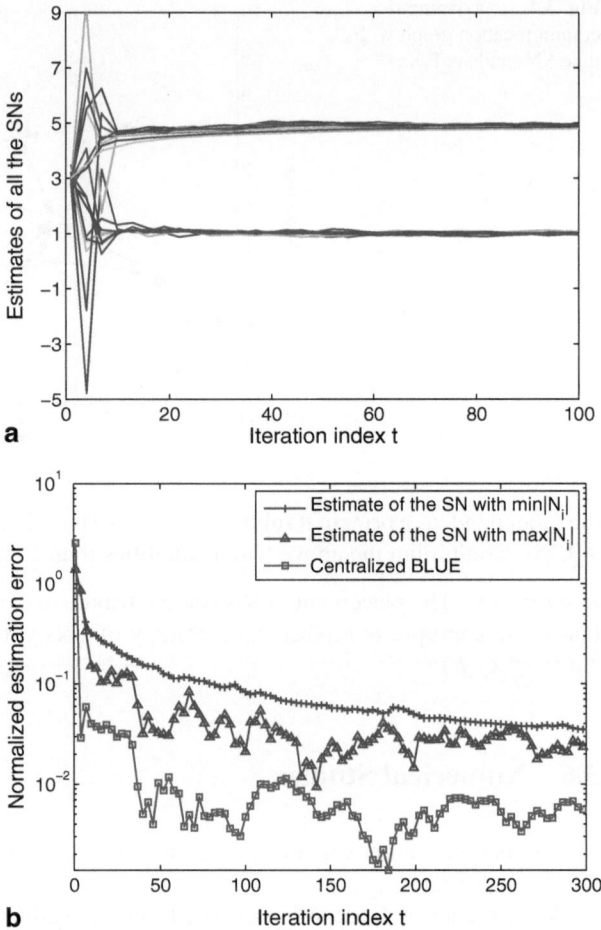

and $x_{i2}(0)$ are both randomly chosen from the interval $[0, 6]$. The following results are averaged over 200 independent experiments.

In Fig. 3.5a, we plot the local estimates of all SNs. Clearly, all the estimates converge to the true value θ. To gain insight, we use the normalized error $e_{\text{norm}i}(t) := \frac{1}{J}\|x_i(t) - \theta\|$ as an indicator. The $e_{\text{norm}i}(t)$'s of SNs with minimum neighbors $\min_i |\mathcal{N}_i|$ and maximum neighbors $\max_i |\mathcal{N}_i|$ are shown in Fig. 3.5b. The normalized estimation error of the centralized estimate x_{BLUE} is also plotted as the benchmark. For the centralized scheme, we use the best linear unbiased estimation (BLUE), which is a minimum variance estimator [76],

$$x_{\text{BLUE}}(t) := \frac{1}{t}\sum_{s=0}^{t-1}\left(\sum_{i=1}^{M}\sigma_i^{-1}H_i^T H_i\right)^{-1}\sum_{i=1}^{M}\sigma_i^{-1}H_i^T y_i(s).$$

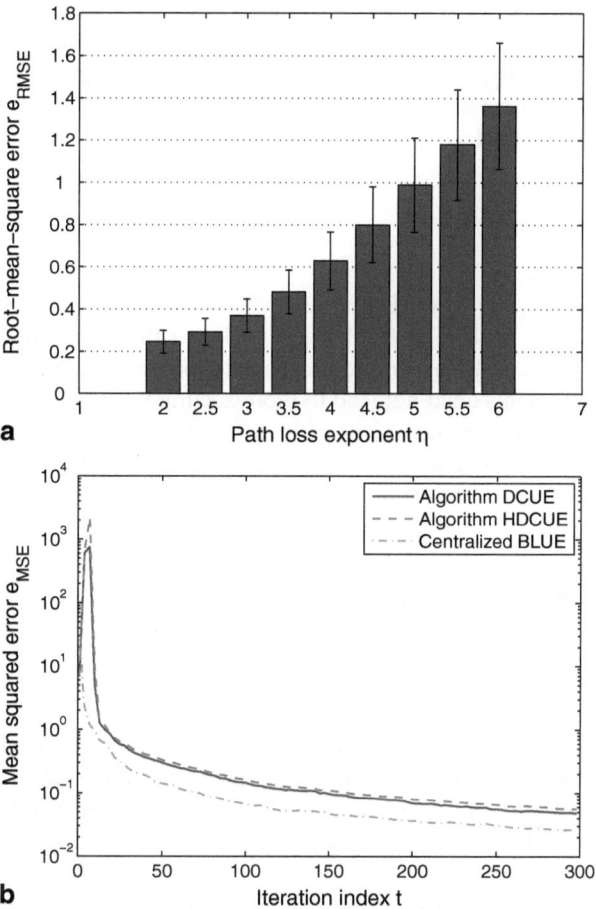

Fig. 3.6 **a** Effect of the path loss exponent η on RMSE. **b** Comparison results of the DCUE algorithm with its homogeneous counterpart HDCUE and the centralized scheme

Figure 3.5b demonstrates that $e_{\mathrm{norm}i}(t)$ of the SN with maximum neighbors is smaller than that of the SN with minimum neighbors and both possess similar behaviors.

To investigate the error decay of the DCUE algorithm, we consider the root-mean-square error (RMSE) over all the SNs e_{RMSE} with the following form:

$$e_{\mathrm{RMSE}} := \frac{1}{MJ} \sum_{i=1}^{M} \sqrt{\frac{1}{K - [K/2] + 1} \sum_{t=[K/2]}^{K} \|e_i(t)\|^2},$$

where K is the total number of iterations of the simulations and $[K/2]$ is the nearest integer around $K/2$. The objective that we only take the summation over the last half of iterations is to avoid the undesired impact of transient behavior on the estimation errors, see Fig. 3.5a. The RMSE versus the value of path loss exponent η is depicted in Fig. 3.6a, which shows that the DCUE algorithm solves the parameter estimation problem for η varied from 2 to 6. And the RMSE increases with the increasing η, which means that the path loss has a great influence on the estimation accuracy.

Finally, we compare our DCUE algorithm with its homogeneous counterpart, i.e., distributed algorithm (3.3) with all 15 nodes being SNs, denoted by HDCUE, and the centralized scheme, with regard to the mean squared error (MSE)

$$e_{\mathrm{MSE}}(t) := \frac{1}{MJ} \sum_{i=1}^{M} \|x_i(t) - \theta\|^2.$$

The result is shown in Fig. 3.6b. It is observed that although there are RNs, the performance of the DCUE algorithm is better than the homogeneous algorithm HDCUE. This reveals that in the presence of asymmetric communication, using more measurements need not lead to more accurate estimates, especially when there are root SNs, the variance of the estimation error of the unobserved variable at the root SNs can in fact increase. This phenomenon has also been observed in [89] for distributed time synchronization.

3.7 Summary

We consider the parameter estimation problem in the relay assisted sensor networks composed of SNs and RNs, where SNs take responsibility for filtering and noise attenuating of the measurements, while RNs only account for data aggregation. A distributed estimation algorithm DCUE is proposed. By using algebraic graph theory and Markov chain approach, the relaying feature of RNs is characterized. Moreover, it is shown that the DCUE algorithm has guaranteed properties of asymptotic unbiasedness and consistency. A quantitative bound on the rate of convergence in the almost sure sense is also established. Finally, simulation results are given to validate the effectiveness of the DCUE algorithm.

Chapter 4
Consensus-Based Optimal Target Tracking in Relay Assisted Wireless Sensor Networks

This chapter is concerned with the problem of filter design for target tracking over WSNs. Similar to Chap. 3, we also consider the relay assisted WSNs with SNs and RNs respectively. In Chap. 3, we deal with the parameter estimation in such networks. However, questions of how to deal with the heterogeneity of nodes and how to design filters for target tracking over such kind of networks remain largely unexplored. We propose in this chapter a novel distributed consensus filter to solve the target tracking problem. Two criteria, namely, unbiasedness and optimality are imposed for the filter design. The so-called sequential design scheme is then presented to tackle the heterogeneity of SNs and RNs. The minimum principle of Pontryagin is adopted for SNs to optimize the estimation errors. As for RNs, the Lagrange multiplier method coupled with the generalized inverse of matrices is then used for filter optimization. Furthermore, it is proven that convergence property is guaranteed for the proposed consensus filter in the presence of process and measurement noise. Simulation results have validated the performance of the proposed filter. It is also demonstrated that the relay assisted WSNs with the proposed filter outperform the homogeneous counterparts in light of reduction of the network cost with slight degradation of estimation performance.

4.1 Introduction

To locate and track a moving target is crucial for many applications such as robotics, surveillance, monitoring, and security for large-scale complex environments [90–92]. In such scenarios, a number of nodes can be employed in order to improve the tracking accuracy and increase the size of the surveillance area in a cooperative manner. Basically, these nodes have modest capabilities of sensing, computation, and multi-hop wireless communication. Equipped with these capabilities, the nodes can self-organize to form a network that are capable of sensing and processing spatial and temporally dense data in the monitored area.

However, due to radio wave propagation loss in wireless channels, reliable communications between nodes can only be ensured within a short distance. As a result, one of the main challenges for target tracking over WSNs is that only local

© The Author(s) 2014   41
C. Chen et al., *Wireless Sensor Networks,*
SpringerBriefs in Computer Science, DOI 10.1007/978-3-319-12379-0_4

information from neighboring nodes is available at each node. In this case, the traditional centralized methods (e.g., Kalman filter and extended Kalman filter [90]) for target tracking are not applicable. One possible way is to decentralize the task over the whole network, which brings forward distributed algorithms. The major advantages of distributed algorithms are that: (i) they can reduce the cost of transmitting all data to the fusion center and (ii) they exhibit resilience against node failures thus achieving robustness of the network. A decentralized version of Kalman filter was proposed in [93], whose main idea is to decentralize the Kalman filtering task to each node. However, this algorithm is not scalable with the network size since each node needs to communicate with all others to compute its local estimate. A more general setting of the distributed Kalman filter was presented in [94], where proximity graph is considered, i.e., each node only needs to share its information with neighboring nodes. Additionally, a diffusion step is added after the estimate update step to improve the performance of distributed Kalman filer.

Recently, consensus algorithms of multi-agent systems have been proven to be effective for undertaking network-wide distributed tasks [88, 95]. Motivated by these algorithms, Olfati-Saber introduced a scalable distributed Kalman-Consensus filter. It is modified from the classic Kalman filer by inserting a consensus term to reduce the disagreements of estimates among the nodes. However, since it is far from optimal with respect to the error covariance, it still might cause unacceptable estimation error. The optimality and stability analysis in its discrete-time form were further studied. An alternative distributed Kalman filer based on consensus algorithms was proposed, which is composed of two stages, i.e., a Kalman-like measurement update and an estimate fusion using consensus matrix. But only scalar system was considered and the optimization of Kalman gain and consensus matrix is in general nonconvex. In [96], a pinning observer was designed to solve the filtering problem in the case that nodes can only observe partial target information. The distributed filtering with partial information between nodes was investigated in [97].

Optimization approach was also used to solve distributed estimation problems. For example, the centralized Kalman filter and smoother for distributed operation through the alternating direction method of multipliers were reformulated. The proposed algorithm offers anytime optimal state estimates based only on local information.

It is noted that most of the aforementioned works deal with homogeneous WSNs, i.e., all nodes possess identical communication and computation capability and so on. In this chapter, we deal with the target tracking problem in a heterogeneous framework. Two types of nodes different on processing abilities, denoted as SNs and RNs, are present in the network. Chapter 3 deals with the parameter estimation in such networks. However, questions of how to deal with the heterogeneity of nodes to facilitate the filter design remain largely unexplored. Our main objective is to propose a distributed optimal consensus filter for target tracking relying only on neighbor-to-neighbor communication among nodes. The heterogeneity of nodes takes new challenges for the optimal filter design as follows: (1) The coexistence of two types of nodes in the network needs novel filter forms for different types of nodes. This makes it impossible to design the optimal filter in a unified framework with identical expressions; (2) The convergence is much more difficult to analyze due to the quite different stochastic representations of the target model and the heterogeneous filter models. The analysis is even intractable without delicate design of the filter. This chapter is based on [98].

The main contributions of this chapter are twofold:

- We propose a distributed consensus filter for target tracking in relay assisted WSNs. The filter parameters are designed according to the performance requirements on unbiasedness and optimality. A sufficient and necessary condition to guarantee the unbiasedness is established. Unlike the homogeneous scenario considered, we develop the so-called sequential design approach to achieve the optimality with node heterogeneity. Specifically, for SNs, the optimization problem is first casted into an optimal control problem. Then the minimum principle of Pontryagin is applied to solve this problem. Then for RNs, Lagrange multiplier and least-square method is used to obtain the optimal weights. Moreover, the discrete-time distributed optimal consensus filter (DOCF) is proposed by discretization and approximation.
- The convergence property of the proposed filter is investigated. We analyze the estimation error dynamics via subtly rewriting the system in the Itô stochastic framework. By using stochastic Lyapunov method and stopping time technique, we show that the estimation error of the consensus filter is exponentially bounded in mean square. Simulations have validated the theoretical analysis.

4.2 Problem Formulation

4.2.1 Target and Network Models

The system considered for target tracking problem in this chapter is composed of N nodes and a moving target in a monitored field. The purpose of the nodes is to cooperatively trace the behavior of the target-based only on neighbor-to-neighbor communication.

Consider a moving target with the following dynamics

$$\dot{x}(t) = Ax(t) + Bw(t), \tag{4.1}$$

where $A \in \mathbb{R}^{n \times n}$ and $B \in \mathbb{R}^{n \times m_1}$ are constant matrices, $x(t) \in \mathbb{R}^n$ is the state vector and $w(t) \in \mathbb{R}^{m_1}$ is the ambient noise, which is assumed to be zero-mean and white. The initial state x_0 is a Gaussian random variable with known mean $\mathbb{E}\{x_0\} = \bar{x}_0$ and positive definite covariance $\mathbb{E}\{(x_0 - \bar{x}_0)(x_0 - \bar{x}_0)^T\} = \Pi_0 > 0$.

The SNs and RNs are deployed in the monitored field to locate and track the moving target. It is assumed that only SNs can sense the environment and observe the target with noisy measurements of the target,

$$y_i(t) = C_i x(t) + v_i(t), \ \forall i \in \mathcal{I}, \tag{4.2}$$

where $C_i \in \mathbb{R}^{n \times m_2}$ is observation matrix and $v_i(t) \in \mathbb{R}^{m_2}$ is the zero-mean and white measurement noise. As for RNs, they can not measure the state of the target, which means that all the information about the target at each RN is directly or indirectly obtained from the SNs.

The nodes in the network are coupled by an undirected communication topology $\mathcal{G} = (\mathcal{V}, \mathcal{E})$, where $\mathcal{V} = \mathcal{I} \cup \mathcal{I}^c = \{1, 2, \ldots, N\}$ is the set of nodes, $\mathcal{I}$ and $\mathcal{I}^c$ are the set of SNs and RNs, respectively, and $\mathcal{E} \subset \mathcal{V} \times \mathcal{V}$ is the set of communication links as in Chap. 3. In this chapter, we also assume that each node transmits at a constant power p_T and the receiver has an ambient noise power n_R. Since path loss is inevitably encountered as the radio wave propagates through the environments, signal power decreases with distance captured by the path loss exponent. Then the signal transmitted from the node i can be successfully received by the node j only if the signal-to-noise ratio (SNR) satisfies

$$\text{SNR} := \frac{p_T}{n_R d_{ij}^{\eta}} \geq \rho,$$

where d_{ij} is the distance between the nodes i and j, η is the path loss exponent (typically, $2 \leq \eta \leq 5$) and $\rho > 0$ is the threshold. Thus, reliable wireless communications between nodes can only be ensured within a distance of $r := \sqrt[\eta]{p_T/(n_R\rho)}$. In this way, the link $(i, j) \in \mathcal{E}$ exists if and only if $d_{ij} \leq r$. Moreover, the set of neighbors of the node i are $\mathcal{N}_i := \{j \in \mathcal{V} : d_{ij} \leq r\}$, $\forall i \in \mathcal{V}$ as defined in Chap. 3.

4.2.2 Distributed Consensus Filter for Relay Assisted WSNs

As it is mentioned, Kalman filter and its distributed generalizations for WSNs [92, 93] are not applicable to target tracking in relay assisted WSNs.

In this section, the method in [99] is used to develop a novel distributed optimal consensus filter for the target tracking scenario. The distributed consensus filter applicable to the relay assisted WSNs is given as follows. For SNs $i \in \mathcal{I}$,

$$\dot{\hat{x}}_i(t) = F_i(t)\hat{x}_i(t) + G_i(t)y_i(t) + H_i(t) \sum_{j \in \mathcal{N}_i} a_{ij} \left[\hat{x}_j(t) - \hat{x}_i(t) \right], \qquad (4.3)$$

and for a RN $i \in \mathcal{I}^c$,

$$\hat{x}_i(t) = \sum_{j \in \mathcal{N}_i} \gamma_{ij}(t)\hat{x}_j(t), \qquad (4.4)$$

where $\hat{x}_i(t) \in \mathbb{R}^n$ is the estimate of the target state $x(t)$, $F_i \in \mathbb{R}^{n \times n}$, $G_i \in \mathbb{R}^{n \times m_2}$ are the filter matrices, $H_i \in \mathbb{R}^{n \times n}$ is the consensus gain matrix, $a_{ij} = \sqrt{p_T/d_{ij}^{\eta}}$ represents the received signal strength at the node i transmitted from the node j, and $\gamma_{ij} > 0$ are constant weights.

Equations (4.3) and (4.4) constitute the distributed consensus filter. In the evolution, the nodes sense and measure the location of the target. Meanwhile, they receive the local estimates from their neighbors. With this information, it is possible for the nodes to update their own estimates via (4.3). Then the nodes transmit the local

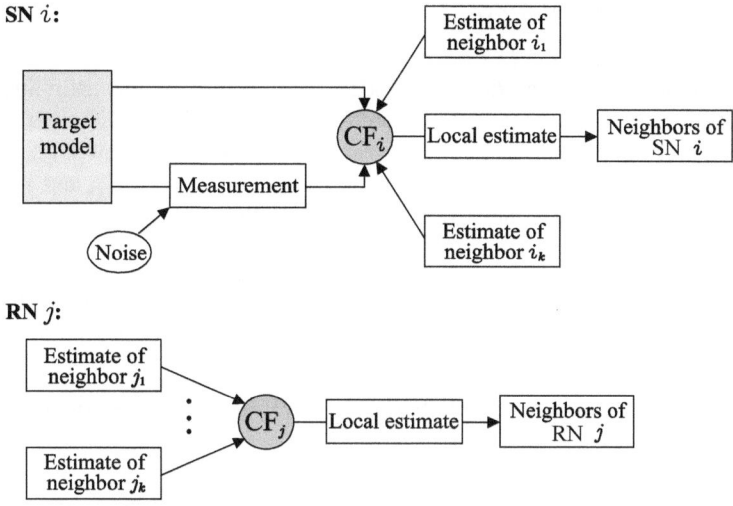

SN i:

RN j:

Fig. 4.1 Schematic representation of the distributed consensus filter for relay assisted WSNs at the node level, where CF is short for consensus filter

estimates to their neighbors. After receiving the local estimates, the RNs scale all the received data in order to find the weighted average in (4.4). Note that in the above process, communication only takes place among neighboring nodes, so the consensus filter (4.3) and (4.4) is totally distributed and thus scalable with the network size. Figure 4.1 depicts a schematic view of the distributed consensus filter at the node level.

Remark 4.1 From (4.3) and (4.4), it can be clearly seen that SNs have more powerful computational ability and thus they can perform more complicated tasks, such as filtering of its measurement and the local estimates from its neighbors. While the abilities of RNs are much limited, they fuse the data to form their own estimates by only simply weighting all the incoming data.

As for the distributed consensus filter (4.3) and (4.4), there are three sets of parameters to be determined, namely, F_i, G_i and γ_{ij}, $i, j \in \mathcal{V}$. To tackle the heterogeneity of (4.3) and (4.4), we propose the sequential design approach to determine these parameters in the next section in light of unbiasedness and optimality of the estimates.

4.3 Structure of Consensus Filter: Unbiasedness and Optimality

In this section, we address the problem of determination of filter matrices $F_i \in \mathbb{R}^{n \times n}$, $G_i \in \mathbb{R}^{n \times m_2}$ and weights $\gamma_{ij} \in \mathbb{R}$ such that the distributed consensus filter (4.3) and (4.4) satisfies the following requirements:

1. $\hat{x}_i(t)$ is an unbiased estimate of $x(t)$, and
2. $\hat{x}_i(t)$ is optimal with respect to a given cost function, for all $i = 1, 2, \ldots, N$.

4.3.1   Unbiased Estimate

The requirement of unbiasedness is widely considered for the estimation. It is not purely imposed for the sake of theoretical derivation, but in some engineering applications, it has been shown to be desirable from a practical viewpoint [76, 100].

The next lemma presents a way to choose the parameters F_i, G_i, and γ_{ij} to meet the requirement of unbiasedness, i.e., $\mathbb{E}\{e_i(t)\} = 0$, $\forall i \in \mathcal{V}, t \geq 0$, where $e_i(t)$ is the estimation error $e_i(t) := \hat{x}_i(t) - x_i(t)$.

Lemma 4.1 *For each $i \in \mathcal{V}$, let the initial estimate $\hat{x}_i(0) = \bar{x}_0$, then $\hat{x}_i(t), \forall t > 0$ is unbiased if and only if (i) for SN, $F_i(t) = A - G_i(t)C_i$ and (ii) for RN, $\sum_{j \in \mathcal{N}_i} \gamma_{ij}(t) = 1$.*

Proof Necessity: Subtracting (4.1) from (4.3) and using (4.4), one obtains for each SN $i \in \mathcal{I}$,

$$
\begin{aligned}
\dot{e}_i(t) = {} & F_i(t)e_i(t) + (F_i(t) + G_i(t)C_i - A)x(t) + G_i(t)v_i(t) \\
& + H_i(t) \sum_{j \in \mathcal{N}_i} a_{ij}\left[e_j(t) - e_i(t)\right] - Bw(t),
\end{aligned}
\tag{4.5}
$$

and for RN $i \in \mathcal{I}^c$

$$
e_i(t) = \sum_{j \in \mathcal{N}_i} \gamma_{ij}(t)e_j(t) + \left(\sum_{j \in \mathcal{N}_i} \gamma_{ij}(t) - 1\right)x(t).
\tag{4.6}
$$

The requirement of unbiasedness means that $\mathbb{E}\{e_i(t)\} = 0$, $\forall i \in \mathcal{V}$. Noting that $\mathbb{E}\{v_i(t)\} = 0$, $\forall i \in \mathcal{I}$ and $\mathbb{E}\{w(t)\} = 0$. Thus by taking expectation on both sides of (4.5) and (4.6), we have $[F_i(t) + G_i(t)C_i - A]\mathbb{E}\{x(t)\} = 0$ and $(\sum_{j \in \mathcal{N}_i} \gamma_{ij}(t) - 1)\mathbb{E}\{x(t)\} = 0$. But, in general, $\mathbb{E}\{x(t)\} \not\equiv 0$, it is necessary that $F_i(t) = A - G_i(t)C_i$, $\forall i \in \mathcal{I}$ and $\sum_{j \in \mathcal{N}_i} \gamma_{ij}(t) = 1$, $\forall i \in \mathcal{I}^c$.

Sufficiency: Imposing the assumptions and then taking expectation on both sides of (4.5), we can obtain

$$
\dot{\bar{e}}_i(t) = F_i(t)\bar{e}_i(t) + H_i(t) \sum_{j \in \mathcal{N}_i} a_{ij}\left[\bar{e}_j(t) - \bar{e}_i(t)\right],
\tag{4.7}
$$

where $\bar{e}_i(t) = \mathbb{E}\{e_i(t)\}$, for all $i \in \mathcal{V}$. Since $\sum_{j \in \mathcal{N}_i} \gamma_{ij}(t) = 1$, it follows from (4.6) and Proposition 1 of [99] that

$$
\bar{e}_i(t) = \sum_{j \in \tilde{\mathcal{I}}} \tilde{\gamma}_{ij}(t)\bar{e}_j(t),
\tag{4.8}
$$

where, $\tilde{\gamma}_{ij}$ satisfy $\sum_{j \in \tilde{\mathcal{I}}} \tilde{\gamma}_{ij}(t) = 1$. Let $\bar{e}(t) = [\bar{e}_1(t), \bar{e}_2(t), \dots, \bar{e}_M(t)]^T$, then by substituting (4.8) to (4.7), one gets the linear system $\dot{\bar{e}}(t) = \Psi(t)\bar{e}(t)$, where $\Psi(t)$ is the coefficient matrix. It is known that the solution is given by $\bar{e}(t) = \tilde{\Psi}(t, 0)\bar{e}(0)$, where $\tilde{\Psi}(t, 0)$ is the state transition matrix. By the assumption, we have $\bar{e}(0) = 0$.

As a result, $\bar{e}(t) = 0$, for all $t > 0$. It thus follows from (4.8) that $\bar{e}_i(t) = 0$, $\forall i \in \mathcal{I}^c, t > 0$. Therefore, the estimate $\hat{x}_i(t)$ is unbiased for all $i \in \mathcal{V}$ and $t > 0$. $\square$

As a result, it follows from Lemma 4.1 that the unbiased filter for SN $i \in \mathcal{I}$ is given by

$$\dot{\hat{x}}_i(t) = (A - G_i(t)C_i)\hat{x}_i(t) + G_i(t)y_i(t) + H_i(t) \sum_{j \in \mathcal{N}_i} a_{ij}[\hat{x}_j(t) - \hat{x}_i(t)]. \quad (4.9)$$

As for RN, from Lemma 3.1 and Lemma 4.1, the unbiased filter (4.4) has the following form

$$\hat{x}_i(t) = \sum_{j \in \tilde{\mathcal{I}}} \tilde{\gamma}_{ij}(t)\hat{x}_j(t), \ i \in \mathcal{I}^c, \quad (4.10)$$

where $\tilde{\mathcal{I}} = \{i \in \mathcal{I} : \text{there exists } j \in \mathcal{I}^c \text{ such that } (i, j) \in \mathcal{E}\}$ and we can rearrange the SNs such that $j_k : \tilde{\mathcal{I}} \to \tilde{\mathcal{I}}$,

$$\tilde{\gamma}_{ij_k}(t) \begin{cases} > 0, & k = 1, \ldots, p_i, \\ = 0, & k = p_i + 1, \ldots, |\tilde{\mathcal{I}}|, \end{cases} \text{ and } \sum_{k=1}^{p_i} \tilde{\gamma}_{ij_k}(t) = 1. \quad (4.11)$$

4.3.2   Optimization of the Gain Matrix G_i and Weights $\tilde{\gamma}_{ij}$

Optimal filters possess some important merits, such as robust in their maintenance of performance standards [100]. In the target tracking scenario, optimal filters for homogeneous sensor networks were considered. In our setting, however, heterogeneity of nodes presents a challenge for the optimal filter design. We propose the sequential design approach to tackle this difficulty.

To facilitate the filter design, we give some standard assumptions. The random variables x_0, $w(t)$, and $v_i(t)$, $i \in \mathcal{I}$, are independent, and

$$\mathbb{E}\{w(t)w^T(\tau)\} = Q(t)\delta(t - \tau), \quad (4.12)$$

$$\mathbb{E}\{v_i(t)v_j^T(\tau)\} = R_{ij}(t)\delta(t - \tau), \ \forall i, j \in \mathcal{I}, \quad (4.13)$$

where $\delta(\cdot)$ is the Dirac delta function. Moreover, assume that $R_i(t) := R_{ii}(t), \forall i \in \mathcal{I}$ is positive definite.

Define the error covariance matrix as $P_{ij}(t) := \mathbb{E}\{e_i(t)e_j^T(t)\}$, for each pair $(e_i(t), e_j(t))$, $i, j \in \mathcal{V}$ and denote $P_{\mathcal{N}_i,j}(t) := \sum_{r \in \mathcal{N}_i} a_{ir}[P_{rj}(t) - P_{ij}(t)]$ and $P_{i,\mathcal{N}_j}(t) := \sum_{s \in \mathcal{N}_j} a_{js}[P_{is}(t) - P_{ij}(t)]$. Then a standard argument shows that $P_{ij}(t)$ satisfies the following cases:

Case i) $i \in \mathcal{I}$ and $j \in \mathcal{I}$.

$$\begin{aligned} \dot{P}_{ij}(t) = & (A - G_i(t)C_i)P_{ij}(t) + P_{ij}(t)(A - G_j(t)C_j)^T \\ & + H_i(t)P_{\mathcal{N}_i,j}(t) + P_{i,\mathcal{N}_j}(t)H_j^T(t) \\ & + G_i(t)R_{ij}(t)G_j^T(t) + BQ(t)B^T; \end{aligned} \quad (4.14)$$

Case ii) $i \in \mathcal{I}$ and $j \in \mathcal{I}^c$.

$$
\begin{aligned}
\dot{P}_{ij}(t) = {} & (A - G_i(t)C_i)P_{ij}(t) + \sum_{k \in \tilde{\mathcal{I}}} \tilde{\gamma}_{jk}(t)P_{ik}(t)(A - G_k(t)C_k)^T \\
& + H_i(t)P_{\mathcal{N}_i,j}(t) + \sum_{k \in \tilde{\mathcal{I}}} \tilde{\gamma}_{jk}(t)P_{i\mathcal{N}_k}(t)H_k^T(t) \\
& + \sum_{k \in \tilde{\mathcal{I}}} \tilde{\gamma}_{jk}(t)G_i(t)R_{ik}(t)G_k^T(t) + BQ(t)B^T;
\end{aligned}
\tag{4.15}
$$

Case iii) $i \in \mathcal{I}^c$ and $j \in \mathcal{I}$.

$$
\begin{aligned}
\dot{P}_{ij}(t) = {} & P_{ij}(t)(A - G_j(t)C_j)^T + \sum_{k \in \tilde{\mathcal{I}}} \tilde{\gamma}_{ik}(t)(A - G_k(t)C_k)P_{kj}(t) \\
& + P_{i\mathcal{N}_j}(t)H_j^T(t) + \sum_{k \in \tilde{\mathcal{I}}} \tilde{\gamma}_{ik}(t)H_k(t)P_{\mathcal{N}_k,j}(t) \\
& + \sum_{k \in \tilde{\mathcal{I}}} \tilde{\gamma}_{ik}(t)G_k(t)R_{kj}(t)G_j^T(t) + BQ(t)B^T;
\end{aligned}
\tag{4.16}
$$

Case iv) $i \in \mathcal{I}^c$ and $j \in \mathcal{I}^c$.

$$
P_{ij}(t) = \sum_{k \in \tilde{\mathcal{I}}} \sum_{l \in \tilde{\mathcal{I}}} \tilde{\gamma}_{ik} \tilde{\gamma}_{jl} P_{kl}(t).
\tag{4.17}
$$

Proof   Under the unbiased requirement, the estimation error $e_i(t)$ satisfies

$$
\dot{e}_i(t) = (A - G_i(t)C_i)e_i(t) + H_i(t) \sum_{j \in \mathcal{N}_i} a_{ij} \left[e_j(t) - e_i(t) \right]
$$
$$
+ G_i v_i(t) - Bw(t), \quad \forall i \in \mathcal{I}.
\tag{4.18}
$$

The solution e_i to (4.18) can be expressed as

$$
e_i(t) = \Phi_i(t,0)e_i(0) + \int_0^t \Phi_i(t,\tau)[\psi_i(\tau) + G_i(\tau)v_i(\tau) - Bw(\tau)]\,d\tau,
$$

where $\Phi_i(t,\tau)$ is the state transition matrix corresponding to $A - G_i(t)C_i - H_i(t)\sum_{r \in \mathcal{N}_i} a_{ir}$ and $\psi_i(\tau) := H_i(\tau)\sum_{r \in \mathcal{N}_i} a_{ir}e_r(\tau)$. Since $x(0)$, $w(t)$ and $v_i(t)$, $\forall i \in \mathcal{I}$ are independent, we can derive that

$$
\mathbb{E}\{e_i(t)v_j^T(t)\} = \underbrace{\int_0^t \Phi_i(t,\tau)\mathbb{E}\{\psi_i(\tau)v_j^T(t)\}d\tau}_{S_1} + \underbrace{\int_0^t \Phi_i(t,\tau)G_i(\tau)\mathbb{E}\{v_i(\tau)v_j^T(t)\}d\tau}_{S_2},
$$

and

$$
\mathbb{E}\{e_i(t)w^T(t)\} = \underbrace{\int_0^t \Phi_i(t,\tau)\mathbb{E}\{\psi_i(\tau)w^T(t)\}d\tau}_{S_3} - \underbrace{\int_0^t \Phi_i(t,\tau)B\mathbb{E}\{w(\tau)w^T(t)\}d\tau}_{S_4}.
$$

Observe that for each $r \in \mathcal{N}_i$, $e_r(\tau)$ only depends on $\{e_i(0), v_i(\tau'), w(\tau'), \tau' \le \tau < t, i \in \mathcal{I}\}$, we find that $S_1 = 0$ and $S_3 = 0$. As for S_2 and S_4, it follows from

(4.12), (4.13), properties of the Dirac delta function and the fact $\Phi_i(t,t) = I_n$ that $S_2 = \frac{1}{2}G_i(t)R_{ij}(t)$ and $S_4 = \frac{1}{2}BQ(t)$. Hence we have

$$\mathbb{E}\{e_i(t)v_j^T(t)\} = \frac{1}{2}G_i(t)R_{ij}(t) \text{ and } \mathbb{E}\{e_i(t)w^T(t)\} = -\frac{1}{2}BQ(t).$$

Following a similar line, one can obtain

$$\mathbb{E}\{v_i(t)e_j^T(t)\} = \frac{1}{2}R_{ij}(t)G_j^T(t) \text{ and } \mathbb{E}\{w(t)e_j^T(t)\} = -\frac{1}{2}Q(t)B^T.$$

Furthermore, we can derive from (4.18) that

$$\begin{aligned}
\dot{P}_{ij}(t) &= \mathbb{E}\{\dot{e}_i(t)e_j^T(t)\} + \mathbb{E}\{e_i(t)\dot{e}_j^T(t)\} \\
&= (A - G_i(t)C_i)P_{ij}(t) + P_{ij}(t)(A - G_j(t)C_j)^T + H_i(t)P_{\mathcal{N}_i,j}(t) + P_{i,\mathcal{N}_j}H_j^T(t) \\
&\quad + G_i(t)\mathbb{E}\{v_i(t)e_j^T(t)\} + \mathbb{E}\{e_i(t)v_j^T(t)\}G_j^T(t) \\
&\quad - B\mathbb{E}\{w(t)e_j^T(t)\} - \mathbb{E}\{e_i(t)w^T(t)\}B^T,
\end{aligned}$$

which together with the above analysis yields (4.14). We only prove (4.15), since (4.16) can be deduced in a similar manner. The unbiasedness reduces (4.6) to be

$$e_i(t) = \sum_{j \in \tilde{\mathcal{I}}} \tilde{\gamma}_{ij}(t)e_j(t), \ \forall i \in \mathcal{I}^c, \tag{4.19}$$

which implies that for any two $i \in \mathcal{I}$ and $j \in \mathcal{I}^c$ one can write $\dot{P}_{ij}$ as follows

$$\dot{P}_{ij}(t) = \mathbb{E}\{\dot{e}_i(t)e_j^T(t)\} + \sum_{l \in \tilde{\mathcal{I}}} \tilde{\gamma}_{jl}\mathbb{E}\{e_i(t)\dot{e}_l^T(t)\}.$$

Then, repeating the similar arguments as in the derivation of (4.14) and bearing in mind the constraint (4.11), we can obtain (4.15). In view of (4.19), $P_{ij}, \forall i, j \in \mathcal{I}^c$ can be expressed as

$$P_{ij}(t) = \mathbb{E}\left\{ \sum_{h \in \tilde{\mathcal{I}}} \tilde{\gamma}_{ih}e_h(t) \sum_{l \in \tilde{\mathcal{I}}} \tilde{\gamma}_{jl}e_l^T(t) \right\} = \sum_{h \in \tilde{\mathcal{I}}} \sum_{l \in \tilde{\mathcal{I}}} \tilde{\gamma}_{ih}\tilde{\gamma}_{jl}P_{hl}(t).$$

which is just (4.17). $\quad\square$

From a physical point of view, it is desirable to have an unbiased estimate with minimum covariance $P_i := P_{ii}$. In the following, however, we consider the perturbed covariance matrix $\tilde{P}_i$ in view of convergence performance,

$$\begin{aligned}
\dot{\tilde{P}}_i(t) &= (A - G_i(t)C_i)\tilde{P}_i(t) + \tilde{P}_i(t)(A - G_i(t)C_i)^T + H_i(t)\tilde{P}_{i,\mathcal{N}_i}(t)^T \\
&\quad + \tilde{P}_{i,\mathcal{N}_i}(t)H_i(t)^T + G_i(t)R_i(t)G_i(t)^T + BQ(t)B^T + W_i,
\end{aligned} \tag{4.20}$$

where W_i is positive definite, $\tilde{P}_{\mathcal{N}_i,j}(t) := \sum_{r \in \mathcal{N}_i} a_{ir}[\tilde{P}_{rj}(t) - \tilde{P}_{ij}(t)]$, $\tilde{P}_{i,\mathcal{N}_j}(t) := \sum_{s \in \mathcal{N}_j} a_{js}[\tilde{P}_{is}(t) - \tilde{P}_{ij}(t)]$ and $\tilde{P}_{ij}$ for all $i, j \in \mathcal{V}$ are likewise determined as P_{ij} in four cases. Denote the corresponding equations as (4.14$'$), (4.15$'$), (4.16$'$) and (4.17$'$), respectively. Moreover, we set $\tilde{P}_i(0) = \tilde{P}_{ij}(0) = \Pi_0$.

We introduce the following cost function

$$\mathcal{J}_i(\tilde{P}_i) = \mathrm{tr}[\tilde{P}_i(t)], i \in \mathcal{V} \tag{4.21}$$

as the measure of the performance of the filter, where $\mathrm{tr}[\,\cdot\,]$ denotes the trace of a matrix. The smaller the value of $\mathcal{J}_i$, the better the consensus filter (4.9) and (4.10).

It is observed from (4.14)–(4.17) that with the cost function $\mathcal{J}_i$ the matrix G_i can be determined independently from the weights $\tilde{\gamma}_{ij}$. This observation makes it possible to sequentially design the parameters G_i and $\tilde{\gamma}_{ij}$ for SNs and RNs, respectively. We first focus on the selection of G_i for SNs.

Lemma 4.2 *For each SN $i \in \mathcal{I}$, the optimal matrix G_i^* with respect to the cost function $\mathcal{J}_i(\tilde{P}_i)$ is given by $G_i^*(t) = \tilde{P}_i^*(t)C_i^T R_i^{-1}(t)$, where $\tilde{P}_i^*(t)$ is the solution of the matrix equation*

$$\begin{aligned}
\dot{\tilde{P}}_i(t) &= A\tilde{P}_i(t) + \tilde{P}_i(t)A^T - \tilde{P}_i(t)C_i^T R_i^{-1}(t)C_i\tilde{P}_i(t) \\
&\quad + H_i(t)\tilde{P}_{i,\mathcal{N}_i}^T(t) + \tilde{P}_{i,\mathcal{N}_i}(t)H_i^T(t) + BQ(t)B^T + W_i.
\end{aligned} \tag{4.22}$$

Proof In order to compute the optimal gain matrix G_i corresponding to the cost function $\mathcal{J}_i(\tilde{P}_i)$, we first note that this minimization problem under the constraint (4.20) is analogous to the classic optimal control problem, where now $\tilde{P}_i$ can be considered as the state of a system and G_i as the control input. Therefore, the minimum principle of Pontryagin can be used here as in [101] where it was employed to derive the centralized Kalman-Bucy filter.

To do this, let t_f be the terminal time and the cost function becomes $\mathcal{J}_i(\tilde{P}_i) = \mathrm{tr}[\tilde{P}_i(t_f)]$, $i \in \mathcal{I}$. Then the optimal control problem with free final state $\tilde{P}_i(t_f)$ is readily followed by introducing the Hamiltonian function $\mathcal{H}_i(\tilde{P}_i, G_i, \Sigma_i, t) = \mathrm{tr}[\dot{\tilde{P}}_i(t)\Sigma_i^T(t)]$, where Σ_i is an $n \times n$ matrix of Lagrange multipliers.

According to the minimum principle of Pontryagin [102], the optimal matrix $G_i^*(t)$ and the corresponding matrix $\Sigma_i^*(t)$ must satisfy the following conditions

$$-\dot{\Sigma}_i^*(t) = \frac{\partial \mathcal{H}_i}{\partial \tilde{P}_i}(\tilde{P}_i^*(t), G_i^*(t), \Sigma_i^*(t), t), \tag{4.23}$$

$$0 = \frac{\partial \mathcal{H}_i}{\partial G_i}(\tilde{P}_i^*(t), G_i^*(t), \Sigma_i^*(t), t), \tag{4.24}$$

$$\Sigma_i^*(t_f) = \frac{\partial \mathcal{J}_i}{\partial \tilde{P}_i}(\tilde{P}_i^*(t_f)). \tag{4.25}$$

In view of matrix calculus, it can be derived that

$$\frac{\partial \mathcal{H}_i}{\partial \tilde{P}_i} = \left(A - G_i^*(t)C_i - \sum_{r \in \mathcal{N}_i} a_{ir}H_i(t)\right)^T \Sigma_i^*(t)$$

$$+ \Sigma_i^*(t)\Big(A - G_i^*(t)C_i - \sum_{r \in \mathcal{N}_i} a_{ir} H_i(t)\Big), \qquad (4.26)$$

$$\frac{\partial \mathcal{H}_i}{\partial G_i^*} = -\Sigma_i^*(t)\tilde{P}_i^*(t)C_i^T - \Sigma_i^{*T}(t)\tilde{P}_i^*(t)C_i^T$$

$$+ \Sigma_i^*(t)G_i^*(t)R_i(t) + \Sigma_i^{*T}(t)G_i^*(t)R_i(t), \qquad (4.27)$$

$$\frac{\partial \mathcal{J}_i}{\partial \tilde{P}_i} = I_n. \qquad (4.28)$$

From the above equations, one can obtain

$$\dot{\Sigma}_i^*(t) = -\Big(A - G_i^*(t)C_i - \sum_{r \in \mathcal{N}_i} a_{ir} H_i(t)\Big)^T \Sigma_i^*(t)$$

$$- \Sigma_i^*(t)\Big(A - G_i^*(t)C_i - \sum_{r \in \mathcal{N}_i} a_{ir} H_i(t)\Big)$$

with the terminal condition $\Sigma_i^*(t_f) = I_n$. It then follows from Proposition 1.1 of [103] that the matrix $\Sigma_i^*(t) > 0$, $\forall t \geq 0$ and thus nonsingular. In consequence, (4.24) together with (4.27) gives $G_i^*(t)R_i(t) = \tilde{P}_i^*(t)C_i^T$, $\forall t \geq 0$. Since $R_i(t)$ is assumed to be positive definite $R_i(t) > 0$, it implies that the optimal gain matrix G_i^* is given by $G_i^*(t) = \tilde{P}_i^*(t)C_i^T R_i^{-1}(t)$.   □

With the optimal G_i, we proceed to the optimization of weights $\tilde{\gamma}_{ij}$ for the RN i such that $\mathcal{J}_i(\tilde{P}_i)$ is minimized.

Lemma 4.3 *The optimal $\tilde{\gamma}_{ij}^*$ can be obtained by*

$$\tilde{\gamma}_{ij}^* = \begin{cases} 1 - \sum_{k=2}^{p_i} z_{jk}, & j = j_1, \\ z_j, & j = j_2, \dots, j_{p_i}, \\ 0, & j = j_{p_i+1}, \dots, j_{|\tilde{\mathcal{I}}|}, \end{cases} \qquad (4.29)$$

where $z := [z_{j_2}, z_{j_3}, \dots, z_{j_{p_i}}]^T > 0$ is the positive solution of the linear equation $U_i z = V_i$, in which

$$U_i := \begin{bmatrix} U_{j_2 j_2} & \cdots & U_{j_2 j_{p_i}} \\ \vdots & \ddots & \vdots \\ U_{j_{p_i} j_2} & \cdots & U_{j_{p_i} j_{p_i}} \end{bmatrix}, \ V_i := \begin{bmatrix} V_{j_2} \\ \vdots \\ V_{j_{p_i}} \end{bmatrix}$$

and $U_{jk jl} := tr[\tilde{P}_{jk jl} - \tilde{P}_{j_l j_1} - \tilde{P}_{jk j_1} + \tilde{P}_{j_1 j_1}]$, $V_{jk} := tr[\tilde{P}_{j_1 j_1} - \tilde{P}_{jk j_1}]$, $k, l = 2, 3, \dots, p_i$.

Proof Note that $\tilde{\gamma}_{ij}$s are constrained by (4.11). By setting $j = i$ in (4.17'), one can collect (4.17') and (4.11) into the following optimization problem to determine the optimal $\tilde{\gamma}_{ij}^*$,

$$
\min_{\substack{\tilde{\gamma}_{ij_k}, \\ k=1,2,\ldots,p_i}} \quad \sum_{h=1}^{p_i} \sum_{l=1}^{p_i} \tilde{\gamma}_{ij_h} \tilde{\gamma}_{ij_l} \, \mathrm{tr}[\tilde{P}_{j_h j_l}]
$$

$$
\text{s.t.} \quad \sum_{h=1}^{p_i} \tilde{\gamma}_{ij_h} = 1, \tag{4.30}
$$

$$
\tilde{\gamma}_{ij_k} > 0, \ k = 1,2,\ldots,p_i.
$$

Substituting $\tilde{\gamma}_{ij_1} = 1 - \sum_{h=2}^{p_i} \tilde{\gamma}_{ij_h}$ into the objective function of (4.30) results in

$$
\min_{\substack{\tilde{\gamma}_{ii_k}, \\ k=2,3,\ldots,p_i}} \quad \sum_{h=2}^{p_i} \sum_{l=2}^{p_i} \tilde{\gamma}_{ii_h} \tilde{\gamma}_{ii_l} \mathrm{tr}[\tilde{P}_{i_h i_l}] + \left(1 - \sum_{h=2}^{p_i} \tilde{\gamma}_{ii_h}\right) \sum_{l=2}^{p_i} \tilde{\gamma}_{ii_l} \, \mathrm{tr}[\tilde{P}_{i_l i_1}]
$$

$$
+ \left(1 - \sum_{h=2}^{p_i} \tilde{\gamma}_{ii_h}\right)^2 \mathrm{tr}[\tilde{P}_{i_1 i_1}]
$$

$$
\text{s.t.} \quad \tilde{\gamma}_{ii_k} > 0, \ k = 2,3,\ldots,p_i.
$$

The above minimization problem can be solved by the method of Lagrangian multipliers. Write the Kuhn–Tucker conditions, we have

$$
\sum_{l=2}^{p_i} \tilde{\gamma}_{ij_l}^* \mathrm{tr}[\tilde{P}_{j_k j_l} - \tilde{P}_{j_1 j_1}] + \left(1 - \sum_{l=2}^{p_i} \tilde{\gamma}_{ij_l}^*\right) \mathrm{tr}[\tilde{P}_{j_k j_1} - \tilde{P}_{j_1 j_1}] = 0
$$

and $\tilde{\gamma}_{ij_k}^* > 0$, for all $k = 2,3,\ldots,p_i$. Then the compact form can be expressed as

$$
U_i \gamma_i^* = V_i, \ \text{ and } \ \tilde{\gamma}_i^* > 0 \tag{4.31}
$$

where $\gamma_i^* := [\tilde{\gamma}_{ii_2}^*, \tilde{\gamma}_{ii_3}^*, \ldots, \tilde{\gamma}_{ii_{p_i}}^*]^T$. This means that the optimal $\tilde{\gamma}_{ij}^*$ can be obtained by solving the linear Eq. (4.31). The proof is thus completed.

In general, there may exist no positive solution or more than one positive solution of $U_i z = V_i$. In this case, we turn to the least-squares solution by minimizing the norm of the residual,

$$
\min_{z \in \mathbb{R}^{p_i}} \quad \|U_i z - V_i\|,
$$

$$
\text{s.t.} \quad z > 0. \tag{4.32}
$$

Several numerical methods can be used here to solve this problem efficiently such as the active set algorithm [104].

Summarizing Lemma 4.1, Lemma 4.2 and Lemma 4.3, we can finally determine the parameters of the consensus filter at each node as follows.

Theorem 4.1 *Consider a relay assisted WSN with communication topology $\mathcal{G}$ tracking the target with dynamics (4.1). The unbiased distributed consensus filter (4.3) and (4.4) is optimal with respect to the cost function $\mathcal{J}_i(\tilde{P}_i)$ (4.21) with the parameters given in Table 4.1.*

Table 4.1 Parameters for the unbiased distributed consensus filter (4.3) and (4.4)

	SN	RN
Unbiasedness	$F_i(t) = A - G_i(t)C_i$	$\sum_{j \in \hat{\mathcal{I}}} \tilde{\gamma}_{ij} = 1$
Optimality	$G_i^*(t) = \tilde{P}_i^*(t)C_i^T R_i^{-1}(t)$	$\tilde{\gamma}_{ij_1}^* = 1 - \sum_{k=2}^{p_i} z_{jk}$, $\tilde{\gamma}_{ij_k}^* = z_j, \forall k = 2, \ldots, p_i,$ and otherwise 0

4.3.3   Discrete-Time Version of the Consensus Filter: DOCF

In the previous subsections, we have demonstrated how to select the parameters F_i, G_i and γ_{ij} by incorporating two criteria on the estimates, namely, unbiasedness and optimality. For practical implementation, here we give its discrete-time version. First, discretize the target dynamics (4.1) and the measurement (4.2) of SNs (see, for example, [90])

$$x(k) = \Phi x(k-1) + B w_d(k-1)$$

and

$$y_i(k) = C_i x(k) + v_{i,d}(k), \; i \in \mathcal{I},$$

where $\Phi := I_n + \epsilon A$, ϵ is the time step, w_d and $v_{i,d}$ are zero-mean and white noise satisfying

$$\mathbb{E}\{w_d(k)w_d^T(l)\} = \varepsilon Q(k)\delta_{kl} := Q^d(k)\delta_{kl},$$

$$\mathbb{E}\{v_{i,d}(k)v_{j,d}^T(l)\} = \frac{R_{ij}(k)}{\varepsilon}\delta_{kl} := R_{ij}^d(k)\delta_{kl}, \; \forall i, j \in \mathcal{I},$$

where $\delta_{kl} = 1$, if $k = l$, and $\delta_{kl} = 0$, otherwise.

We assume that all nodes in the network are synchronized so that their communication and estimate updates can be concurrently performed. In this way, the discrete-time filter distributed optimal consensus filter (DOCF) is summarized in Algorithm 1. In this algorithm, SNs and RNs are treated separately due to the sequential design approach. For each SN, it first computes the optimal gain G_i, broadcasts it to all its neighbors and then updates its estimates (line 4 and line 5). As for the estimate update, it fuses its measurement and the local estimates received from its neighbors (line 5). Then, it calculates the perturbed error covariance $\tilde{P}_{ij}$ based on the received data from its neighbors (line 6). On the other hand, for each RN, it only incorporates all its neighbors' estimates (line 9) and the perturbed error covariance (line 10) based on the optimal weights $\tilde{\gamma}_{ij}$ (line 8).

Remark 4.2   It is noted that in Algorithm 1, we make some approximations of G_i and $\tilde{P}_{ij}$ by neglecting the $o(\varepsilon)$ and $o(\varepsilon^2)$ terms during discretization, respectively (see line 4 and line 6). Moreover, it requires that at every iteration, SN i should transmit the prior estimate $\hat{x}_i^-$, the corresponding covariance matrix $\tilde{P}_{ij}^-$, the optimal gain G_i

and the measurement matrix C_i to its neighbors and even its second-hop neighbors (see line 6). It might become a heavy burden for the SNs with limited energy. In this case, a suboptimal alternative of the update of $\tilde{P}_{ij}$ could be

---

Algorithm 1 Algorithm DOCF implemented on node i

1: Initialization: $\hat{x}_i^-(0) = \mathbb{E}\{x(0)\}$, $\tilde{P}_i^-(0) = \tilde{P}_{ij}^-(0) = \Pi_0$.
2: **loop** {Local iteration}
3:     **if** $i \in \mathscr{I}$ **then**
4:         Compute the optimal gain

$$G_i(k) = \tilde{P}_i^-(k)C_i^T \left(R_i^d(k) + C_i\tilde{P}_i^-(k)C_i^T\right)^{-1}.$$

5:         Take measurement $y_i(k)$ and update its local estimate

$$\hat{x}_i(k) = \hat{x}_i^-(k) + G_i(k)\left(y_i(k) - C_i\hat{x}_i^-(k)\right) + \varepsilon H_i(k) \sum_{r \in \mathscr{N}_i} a_{ir}\left(\hat{x}_r^-(k) - \hat{x}_i^-(k)\right).$$

6:         Compute the perturbed error covariance

$$\tilde{P}_{ij}(k) = (I_n - G_i(k)C_i)\tilde{P}_{ij}^-(k)(I_n - G_j(k)C_j)^T + \varepsilon H_i(k)\sum_{r \in \mathscr{N}_i} a_{ir}\left(\tilde{P}_{rj}^-(k) - \tilde{P}_{ij}^-(k)\right)$$
$$+ \varepsilon \sum_{s \in \mathscr{N}_j} a_{js}\left(\tilde{P}_{is}^-(k) - \tilde{P}_{ij}^-(k)\right)H_j^T(k) + G_i(k)R_{ij}^d(k)G_j^T(k).$$

7:     **else if** $i \in \mathscr{I}^c$ **then**
8:         Compute the optimal weights $\tilde{\gamma}_{ij}$, $j \in \mathscr{I}$ according to (4.32) and $\tilde{\gamma}_{ii_1} = 1 - \sum_{h=2}^{p_i} \tilde{\gamma}_{ii_h}$.
9:         Fuse the data received from its neighbors

$$\hat{x}_i(k) = \sum_{j \in \mathscr{I}} \tilde{\gamma}_{ij}\hat{x}_j(k).$$

10:        Compute the perturbed error covariance

$$\tilde{P}_{ij}(k) = \sum_{r \in \mathscr{I}} \sum_{s \in \mathscr{I}} \tilde{\gamma}_{ir}\tilde{\gamma}_{js}\tilde{P}_{rs}(k).$$

11:    **end if**
12:    Update the state of the consensus filter

$$\hat{x}_i^-(k) = \Phi\hat{x}_i(k-1),$$

$$\tilde{P}_{ij}^-(k) = \Phi\tilde{P}_{ij}(k-1)\Phi^T + BQ^d(k-1)B^T + \varepsilon W_i.$$

13: **end loop**

$$\tilde{P}_{ij}(k) = (I_n - G_i(k)C_i)\tilde{P}_{ij}^-(k)(I_n - G_j(k)C_j)^T + G_i(k)R_{ij}^d(k)G_j^T(k).$$

4.4   Mean-Square Analysis

In the previous sections, we propose a distributed consensus filter and investigate its properties of unbiasedness and optimality. But these properties provide no clue about its stability and convergence, which is important for the theoretical analysis and practical application. In this section, we give the mean-square analysis of the filter in the Itô stochastic framework.

We first rewrite the target dynamics (4.1) in the form of Itô stochastic differential equation [105]

$$dx(t) = Ax(t)dt + Bd\tilde{w}(t), \qquad (4.33)$$

where $\tilde{w}(t)$ is an m_1-dimensional Brownian motion with $\mathbb{E}\{d\tilde{w}(t)d\tilde{w}^T(t)\} = Q(t)dt$. In order to obtain a tractable mathematical interpretation of the measurement (4.2), we introduce a stochastic process $z_i(t) = \int_0^t y_i(s)ds$, $\forall i \in \mathcal{I}$. Then its stochastic representation is given by

$$dz_i(t) = C_i x(t)dt + d\tilde{v}_i(t), \ \forall i \in \mathcal{I}, \qquad (4.34)$$

where $\tilde{v}_i(t)$ is an m_2-dimensional Brownian motion with $\mathbb{E}\{d\tilde{v}_i(t)d\tilde{v}_j^T(t)\} = R_{ij}(t)dt$, independent of x_0 and $\tilde{w}(t)$.

Accordingly, the optimal consensus filter (4.3) can be rewritten as

$$d\hat{x}_i(t) = A\hat{x}_i(t)dt + G_i^*(t)(dz_i(t) - C_i\hat{x}_i(t)dt)$$
$$+ H_i(t) \sum_{j \in \mathcal{N}_i} a_{ij}[\hat{x}_j(t) - \hat{x}_i(t)]dt, \ \forall i \in \mathcal{I}. \qquad (4.35)$$

Denote $F_i^* := A - G_i^* C_i$, then subtracting (4.33) from (4.35) and using (4.34) leads to the stochastic representation

$$de_i(t) = F_i^*(t)e_i(t)dt + H_i(t) \sum_{j \in \mathcal{N}_i} a_{ij}[e_j(t) - e_i(t)]dt + \begin{bmatrix} -B & G_i^*(t) \end{bmatrix} \begin{bmatrix} d\tilde{w}(t) \\ d\tilde{v}_i(t) \end{bmatrix}.$$
$$(4.36)$$

Stack all estimation errors and noise into vectors $e := [e_1, e_2, \ldots, e_M]^T$ and $\tilde{v} := [\tilde{w}, \tilde{v}_1, \ldots, \tilde{v}_M]^T$, respectively and define matrices $\tilde{P} := [\tilde{P}_{ij}] \in \mathbb{R}^{M \times M}$, $R := [R_{ij}] \in \mathbb{R}^{M \times M}$, and then we can obtain the compact vector form of (4.36) and (4.20) as follows.

Lemma 4.4 *Under the requirement of unbiasedness, the compact forms of the error dynamics (4.36) and the perturbed covariance dynamics (4.20) can be rewritten as*

$$de(t) = \Psi(t)e(t)dt + \Gamma(t)d\tilde{v}(t), \qquad (4.37)$$

and

$$\dot{P}(t) = \Psi(t)\tilde{P}(t) + \tilde{P}(t)\Psi^T(t) + G^*(t)R(t)G^{*T}(t) + (\mathbf{1}_M\mathbf{1}_M^T) \otimes (BQ(t)B^T) + W, \qquad (4.38)$$

where $\Psi = F^* - H(L \otimes I_n)$, $F^* := diag\{F_1, F_2, \ldots, F_M\}$, $H := diag\{H_1, H_2, \ldots, H_M\}$, $G^* := diag\{G_1^*, G_2^*, \ldots, G_M^*\}$, $\Gamma := [-\mathbf{1}_M \otimes B \ \ G^*]$, $W := diag\{W_1, W_2, \ldots, W_M\}$, *and* $L := [l_{ij}] \in \mathbb{R}^{M \times M}$ *with*

$$l_{ij} = \begin{cases} \displaystyle\sum_{j \in \mathcal{N}_i \cap \mathcal{I}} a_{ij} + \sum_{j \in \mathcal{N}_i \cap \mathcal{I}^c} \sum_{k \in \tilde{\mathcal{I}} \setminus \{i\}} a_{ij}\tilde{\gamma}_{jk}, & j = i, \\ -a_{ij} - \displaystyle\sum_{k \in \mathcal{N}_i \cap \mathcal{I}} a_{ik}\tilde{\gamma}_{kj}, & j \neq i. \end{cases}$$

Proof Note that $\tilde{\gamma}_{ij}$ satisfies the condition (4.11). From (4.19), we obtain for all $i \in \mathcal{I}$

$$
\sum_{j \in \mathcal{N}_i} a_{ij}(e_j - e_i) = \sum_{j \in \mathcal{N}_i \cap \mathcal{I}} a_{ij}(e_j - e_i) + \sum_{j \in \mathcal{N}_i \cap \mathcal{I}^c} \sum_{k \in \tilde{\mathcal{I}}} a_{ij}\tilde{\gamma}_{jk}(e_k - e_i)
$$

$$
= \sum_{j \in \mathcal{N}_i \cap \mathcal{I}} a_{ij}(e_j - e_i) + \sum_{j \in \tilde{\mathcal{I}} \setminus \{i\}} \sum_{k \in \mathcal{N}_i \cap \mathcal{I}^c} a_{ik}\tilde{\gamma}_{kj}(e_j - e_i)
$$

$$
= -(l_i^T \otimes I_n)e, \tag{4.39}
$$

where $\otimes$ is the Kronecker product and $l_i := [l_{i1}, l_{i2}, \ldots, l_{iM}]^T$. Consequently, we can collect (4.36) and (4.19) into the compact vector form given in (4.37). Similar to the derivation of (4.39), it is not difficult to derive from ((4.14)′)–((4.17)′) and (4.22) that dynamics of $\tilde{P}$ can be expressed as (4.38). This completes the proof.

In order to investigate the mean-square performance of the error dynamics (4.37) with (4.38), we adopt the following definition of stochastic stability (see, for example, [106, 107]).

Definition 4.1 The stochastic process $e(t)$ is said to be exponentially bounded in mean-square, if there exist constants $\beta_i > 0$, $i = 1, 2, 3$ such that

$$
\mathbb{E}\{\|e(t)\|^2\} \leq \beta_1 \mathbb{E}\{\|e(0)\|^2\} \exp(-\beta_2 t) + \beta_3, \ \forall t \geq 0.
$$

We are now ready to give the following theorem.

Theorem 4.2 *Consider the stochastic differential Eqs. (4.37) and (4.38). Suppose that $\tilde{P}(t)$, $Q(t)$, and $R(t)$ are bounded, i.e., there exist positive constants $\underline{p}, \bar{p}, \bar{q}, \underline{r}, \bar{r} > 0$ such that*

$$
\underline{p} I_{nM} \leq \tilde{P}(t) \leq \bar{p} I_{nM}, \tag{4.40}
$$

$$
Q(t) \leq \bar{q} I_n, \tag{4.41}
$$

$$
\underline{r} I_{mM} \leq R(t) \leq \bar{r} I_{mM}, \tag{4.42}
$$

then the estimation error $e(t)$ is exponentially bounded in mean square.

Proof Define the stochastic process $V(e,t) = e^T \tilde{P}^{-1}(t)e$ corresponding to the estimation error dynamics (4.37) with (4.38). From (4.40), we have

$$
\frac{1}{\bar{p}}\|e\|^2 \leq V(e,t) \leq \frac{1}{\underline{p}}\|e\|^2. \tag{4.43}
$$

Then utilizing the Itô's formula [108], we derive that

$$
dV(e,t) = \mathcal{L}V(e,t)dt + 2e^T \tilde{P}^{-1}(t)\Gamma(t)d\tilde{v}(t), \tag{4.44}
$$

where

$$
\mathcal{L}V(e,t) = -e^T \tilde{P}^{-1}(t)\dot{\tilde{P}}(t)\tilde{P}^{-1}(t) + 2e^T \tilde{P}^{-1}(t)\Psi(t)e \\
+ \mathrm{tr}\left[\Gamma(t) \, \mathrm{diag}\{Q(t), R(t)\}\Gamma^T(t)\tilde{P}^{-1}(t)\right]. \tag{4.45}
$$

With the assumptions (4.40), (4.41), (4.42) and the expression of G_i^*, one obtains

$$\text{tr}\left[\Gamma(t)\,\text{diag}\{Q(t), R(t)\}\Gamma^T(t)\tilde{P}^{-1}(t)\right]$$

$$= \text{tr}\left[((\mathbf{1}_M\mathbf{1}_M^T)\otimes(BQ(t)B^T) + G^*(t)R(t)G^{*T}(t))\tilde{P}^{-1}(t)\right]$$

$$\leq \frac{1}{\underline{p}}\left(nM\bar{q}\lambda_{\max}(BB^T) + \bar{r}\sum_{i=1}^{M}\text{tr}[\tilde{P}_i(t)C_i^T R_i^{-2}C_i\tilde{P}_i(t)]\right)$$

$$\leq \frac{n\vartheta_M}{\underline{p}}, \tag{4.46}$$

where $\vartheta_M = M\bar{q}\lambda_{\max}(BB^T) + \bar{r}\bar{p}^2/\underline{r}^2\sum_{i=1}^{M}\lambda_{\max}(C_i^T C_i)$ and $\lambda_{\max}(\cdot)$ means the largest eigenvalue. As a result, substituting (4.38) and (4.46) into (4.45) leads to

$$\mathcal{L}V(e,t) \leq -e^T\tilde{P}^{-1}(t)W\tilde{P}^{-1}(t)e + \frac{n\vartheta_M}{\underline{p}}, \tag{4.47}$$

where in the last line use was made of positive semidefiniteness of $BQ(t)B^T$ and $G^*(t)R(t)G^{*T}(t)$.

Combining the fact that $W > 0$ is positive definite and relation (4.43) finally enables (4.47) to be

$$\mathcal{L}V(e,t) \leq -\frac{\lambda_{\min}(W)}{\bar{p}^2}\|e\|^2 + \frac{n\vartheta_M}{\underline{p}}$$

$$\leq -\kappa_1 V(e,t) + \kappa_2, \tag{4.48}$$

where $\kappa_1 = \underline{p}\lambda_{\min}(W)/\bar{p}^2$ and $\kappa_2 = n\vartheta_M/\underline{p}$.

To complete the proof, we use the stopping time technique of stochastic differential theory [108]. For any given time $T \geq 0$ and each positive integer $k \geq \mathbb{E}\{\|e(0)\|\}$, define

$$\tau_{k,T} = \begin{cases} \inf\{t \geq 0 : \|e(t)\| \geq k\}, & \text{if } \exists t \in [0,T], \|e(t)\| \geq k, \\ T, & \text{otherwise.} \end{cases}$$

Let $\tau_{k,T}^t := \min\{t, \tau_{k,T}\}$, then it is apparent $\lim_{k\to\infty}\tau_{k,T}^t = t$ almost surely, for all $0 \leq t \leq T$.

By Itô's formula and using (4.48), we have

$$d(\exp(\kappa_1 t)V(e,t)) \leq \kappa_2\exp(\kappa_1 t)dt + 2\exp(\kappa_1 t)e^T\tilde{P}^{-1}(t)\Gamma(t)d\tilde{v}. \tag{4.49}$$

Integrating and then taking expectation of both sides of the above equation, we arrive at the following relations

$$\mathbb{E}\{V(e(\tau_{k,T}^t), \tau_{k,T}^t)\} \leq \exp(-\kappa_1\tau_{k,T}^t)\,\mathbb{E}\{V(e(0),0)\} + \mathbb{E}\left\{\int_0^{\tau_{k,T}^t}\kappa_2\exp(\kappa_1(s-\tau_{k,T}^t))ds\right\}$$

$$+2\mathbb{E}\left\{\int_0^{\tau_{k,T}^t}\exp(\kappa_1(s-\tau_{k,T}^t))e^T(s)\tilde{P}^{-1}(s)\Gamma(s)d\tilde{v}(s)\right\}$$

$$\leq \mathbb{E}\{V(e(0),0)\} + \frac{\kappa_2}{\kappa_1},$$

where in the second inequality use was made of properties of Itô integral [108]. Note that $\lim_{k\to\infty} \tau_{k,T}^t = t$ almost surely, we can now apply Fatou's lemma to the above inequality to obtain

$$
\begin{aligned}
\mathbb{E}\{V(e,t)\} &\leq \lim_{k\to\infty} \mathbb{E}\{V(e(\tau_{k,s}^t), \tau_{k,T}^t)\} \\
&\leq \mathbb{E}\{V(e(0),0)\} + \frac{\kappa_2}{\kappa_1} < \infty, \ \forall t \in [0,T].
\end{aligned}
$$

It thus follows from (4.40)-(4.42), (4.43) and (4.46) that there exists a constant $\mu > 0$ such that

$$
\begin{aligned}
\mathbb{E}\left\{ \int_0^t \|e^T(s)\tilde{P}^{-1}(s)\Gamma(s)\sqrt{\text{diag}\{Q(s),R(s)\}}\|^2 ds \right\} &\leq \mu\bar{p}\, \mathbb{E}\left\{ \int_0^t V(e(s),s)ds \right\} \\
&\leq \mu\bar{p}T\left(\mathbb{E}\{V(e(0),0)\} + \frac{\kappa_2}{\kappa_1} \right) \\
&< \infty, \ \forall t \in [0,T].
\end{aligned}
$$

Since T is arbitrary, the properties of Itô integral [108] yield

$$
\mathbb{E}\left\{ \int_0^t e^T(s)\tilde{P}^{-1}(s)\Gamma(s)d\tilde{v}(s) \right\} = 0, \ \forall t \geq 0.
$$

Combining this with (4.49), we find that

$$
\mathbb{E}\{V(e,t)\} \leq \exp(-\kappa_1 t)\mathbb{E}\{V(e(0),0)\} + \frac{\kappa_2}{\kappa_1}(1 - \exp(\kappa_1 t)),
$$

which together with (4.43) reveals

$$
\mathbb{E}\{\|e(t)\|^2\} \leq \frac{\bar{p}}{\underline{p}}\mathbb{E}\{\|e(0)\|^2\}\exp(-\kappa_1 t) + \frac{\kappa_2\bar{p}}{\kappa_1}.
$$

Therefore, by Definition 4.1, we can conclude that the estimation error $e(t)$ is exponentially bounded in mean square. $\qquad\square$

Remark 4.3 In Theorem 4.2, bounds of $\tilde{P}$ are required to prove the exponential boundedness of the estimation error $e(t)$. The condition (4.40) is closely related with the observability and detectability properties of the linear system (4.33). For more details, please refer to [109].

Remark 4.4 If we only want to obtain the error bounds, $Q(t)$ and $R(t)$ need not to be the covariances of the noise terms. Any positive definite matrices could be applied.

4.5 Simulation Study

In this section, simulations are presented to verify the efficiency of the proposed DOCF algorithm to track a maneuvering target governed by Singer model in 2-D space.

Singer model is widely used for modeling maneuvering target in the 1-D space in the literature of target tracking [90, 110, 111], which assumes that the target acceleration $a(t)$ is modeled as the first-order stationary Markov process

$$\dot{a}(t) = -\alpha a(t) + \omega(t), \ \alpha > 0, \tag{4.50}$$

where $\omega(t)$ is zero-mean white noise with $\mathbb{E}\{\omega(t)\omega(\tau)\} = 2\alpha\sigma_m^2\delta(t-\tau)$ and σ_m^2 is the instantaneous variance of $a(t)$. The state representation of the continuous-time Singer model in the 2-D space can be expressed as

$$\dot{x}(t) = \begin{bmatrix} 0 & 0 & 1 & 0 & 0 & 0 \\ 0 & 0 & 0 & 1 & 0 & 0 \\ 0 & 0 & 0 & 0 & 1 & 0 \\ 0 & 0 & 0 & 0 & 0 & 1 \\ 0 & 0 & 0 & 0 & -\alpha_1 & 0 \\ 0 & 0 & 0 & 0 & 0 & -\alpha_2 \end{bmatrix} x(t) + \begin{bmatrix} 0 & 0 \\ 0 & 0 \\ 0 & 0 \\ 0 & 0 \\ 1 & 0 \\ 0 & 1 \end{bmatrix} \begin{bmatrix} \omega_1(t) \\ \omega_2(t) \end{bmatrix},$$

where $x = [x_1, x_2, x_3, x_4, x_5, x_6]^T$ of which $[x_1, x_2]^T$ is the position vector, $[x_3, x_4]^T$ is the velocity vector, $[x_5, x_6]^T$ is the acceleration vector and $\omega_1(t), \omega_2(t)$ are independent jerk noise along the X-axis and Y-axis, respectively. The initial conditions are $x_0 = [-30, 0, 1, 0.5, 0.2, 0.1]^T$ and $\Pi_0 = I_6$.

Typical values of the parameter $1/\alpha$ for an evasive maneuver are 10-20 sec as suggested in [110]. In the simulations, we choose $\alpha_1 = 0.1$ and $\alpha_2 = 0.05$. And the instantaneous variances $\sigma_{m_1}^2$ and $\sigma_{m_2}^2$ are set to be

$$\sigma_{m_1}^2 = \frac{\left(\sqrt{0.27}\right)^2}{3}[1 + 4 \times 0.2 - 0.3] = 0.135,$$

$$\sigma_{m_2}^2 = \frac{\left(\sqrt{0.54}\right)^2}{3}[1 + 4 \times 0.2 - 0.3] = 0.27,$$

so that $Q = 0.027 I_2$.

We use a sensor network of $N = 20$ nodes consisting of 12 SNs $\mathcal{I} = \{1, 2, \ldots, 12\}$ and 8 RNs $\mathcal{I}^c = \{13, 14, \ldots, 20\}$ as shown in Fig. 4.2 to track the maneuvering target. This network is obtained by distributing the nodes randomly over a squared area of side length 100 m. And any two nodes can communicate if the distance between them is smaller than $r = 30$. Moreover, each SN can observe the distorted position of the target according to the linear model (4.2) with

$$C_i = \begin{bmatrix} 1 & 0 & 0 & 0 & 0 & 0 \\ 0 & 1 & 0 & 0 & 0 & 0 \end{bmatrix}, \ R_i = 0.25\sqrt{d_{it}}I_2, \ \forall i \in \mathcal{I},$$

where d_{it} is the distance between SN i and the target. The factor $\sqrt{d_{it}}$ in R_i means that the farther it is from the target, the less information can be observed by the SN. The

Fig. 4.2 A relay assisted *WSN* of 20 nodes with 12 *SNs* and 8 RNs. The SN 2 is highlighted, which has 3 neighbors, i.e., 2 SN neighbors and 1 RN neighbor

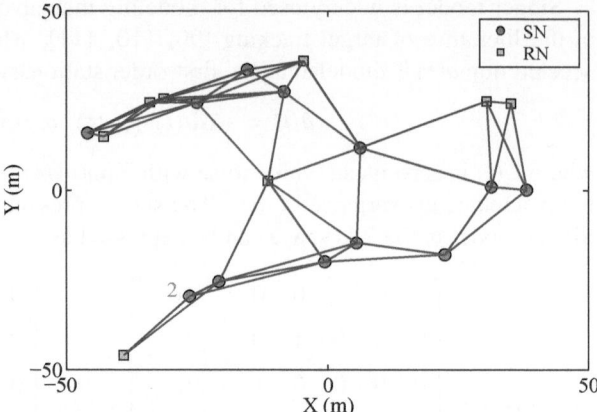

Fig. 4.3 Tracking result of the proposed *DOCF* algorithm: the true and estimated trajectories of the target at the SN 2. The arrows are the error vectors pointing to the true positions. And the rectangles represent the 3σ uncertainty bounds for the target positions centered at the estimates at the corresponding time steps

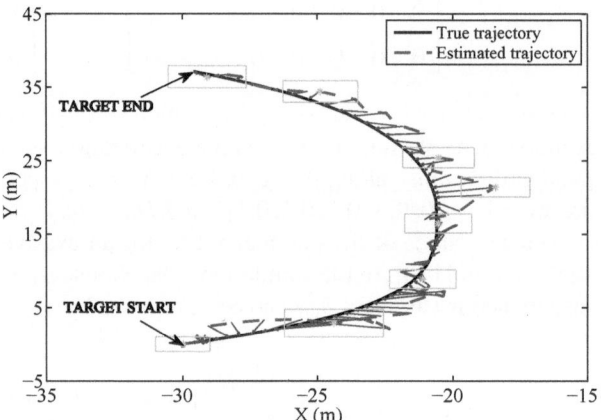

other parameters used in the simulations are: $H_i = I_6$, $\eta = 2$ and $a_{ij} = \sqrt{1/(1 + d_{ij}^{\eta})}$. Each result presented in this section is the average of 20 independent runs except otherwise stated.

Figure 4.3 displays the true and estimated trajectories of the target at the SN 2 in one run, from which we can observe that the estimates are close to the true trajectory. This implies that the proposed DOCF algorithm is able to track the target for SN 2. A common sanity test of the estimates is the consistency testing [90], which is crucial for the optimality evaluation. Figure 4.3 also demonstrates that the proposed DOCF algorithm produces consistent estimates, in other words, the true target positions are almost within the 3σ uncertainty bounds centered at the corresponding estimates. Actually, the simulation results read that up to 91.4 % of x_js fall within the interval $\left[\hat{x}_{2,j} - 3\sqrt{P_2(j,j)}, \ \hat{x}_{2,j} + 3\sqrt{P_2(j,j)} \right]$, $j = 1, 2$.

To qualitatively evaluate the performance of the proposed DOCF algorithm, we introduce the following two metrics, namely, the average mean-square deviation (MSD) over all nodes $\text{MSD} = \frac{1}{N} \sum_{i=1}^{N} \|\hat{x}_i - x\|^2$, and the average disagreement of the

Fig. 4.4 *MSD* performance and *DoE* performance of the proposed DOCF algorithm

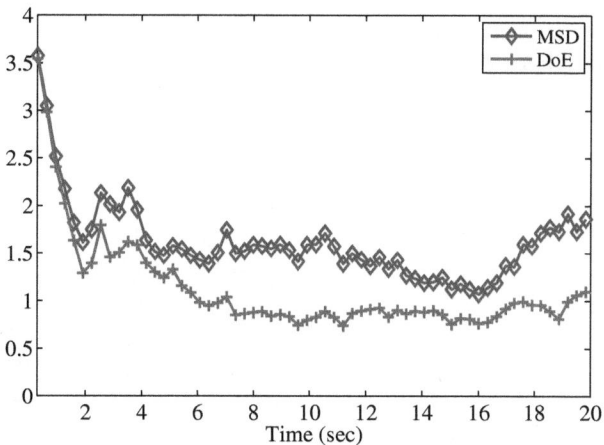

estimates (DoE) among all nodes $\text{DoE} = \frac{1}{N} \sum_{i=1}^{N} \|\hat{x}_i - \text{Ave}(\hat{x})\|^2$, where $\text{Ave}(\hat{x}) = \frac{1}{N} \sum_{i=1}^{N} \hat{x}_i$. Note that MSD measures the estimation accuracy of the proposed DOCF algorithm, while DoE characterizes the differences of the estimates which reflects the degree of consensus among all nodes. We plot the average MSD and DoE in Fig. 4.4. It can be seen that both MSD and DoE are small with respect to the measurement noise, which reveals that the proposed DOCF algorithm can not only track the target with small errors, but also possesses the ability of enabling all the nodes to approximately reach agreement on their estimates.

In order to avoid the undesired impact of transient behavior on the estimation errors, we use the average of the root-mean-square error (RMSE) over all nodes to measure the overall performance of the algorithms, which is defined as

$$\text{RMSE} = \frac{1}{N} \sum_{i=1}^{N} \sqrt{\frac{1}{K - [K/2] + 1} \sum_{k=[K/2]}^{K} \|e_i(k)\|^2},$$

where K is the total number of time steps of simulations and $[K/2]$ is the nearest integer around $K/2$.

Figure 4.5 depicts RMSE-X, RMSE-Y, and RMSE versus the number of SNs, where RMSE-X and RMSE-Y denote the RMSEs for X-position x_1 and Y-position x_2, respectively. The result demonstrates that the differences of RMSE-X, RMSE-Y, and RMSE are small as the number of SNs varies from 4 to 20, respectively, which indicates that the target tracking problem can be solved using relay assisted WSNs. We can further see that the RMSE is the smallest when the number of SNs is 12. This is reasonable because (i) the measurement noise v_i is heavily influenced by the distance from the target d_{it} for each SN i and (ii) only local estimates are communicated between SNs in the DOCF algorithm. If the number of SNs is small, few information about the target could be obtained by the SNs. While if the number of SNs is large, more noise might be injected to the data about the target. Therefore, in both cases,

Fig. 4.5 Effect of the number
of SNs on RMSE-X,
RMSE-Y and RMSE

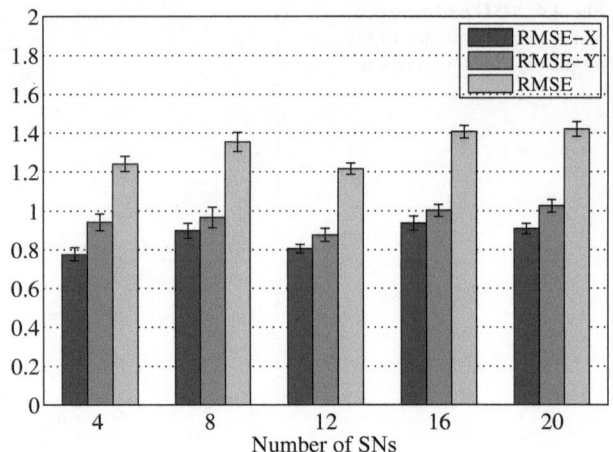

lower RMSE is certainly not expected. And this suggests that there might be an optimal number of SNs, which is one of our future works to find this optimal value.

Finally, we compare the performance of the proposed DOCF algorithm with centralized Kalman fitler, KCF (Algorithm 3 in [96]), diffKF (Algorithm 2 in [94]) and the non-optimal algorithms. In order for fair comparison, we consider the local measurements-based DOCF algorithm denoted as DOCF-M, which is given by

$$\dot{\hat{x}}_i(t) = (A - G_{i,M}(t)C_{i,M})\hat{x}_i(t) + G_{i,M}(t)y_{i,M}(t) + H_i(t) \sum_{j \in \mathcal{N}_i} [\hat{x}_j(t) - \hat{x}_i(t)],$$

where $y_{i,M} = [y_{i_1}^T, \ldots, y_{i_{|\mathcal{N}_i|}}^T]^T$ is the stacked vector of its neighbors' measurements and $C_{i,M} = [C_{i_1}^T, \ldots, C_{i_{|\mathcal{N}_i|}}^T]^T$. It is straightforward to formally derive the optimal consensus filter in this case following the same line as in the previous sections, so we omit the details. For the non-optimal consensus filter, we set $G_i(t) = \tilde{P}_i(t)C_i^T R_i^{-1} + 0.01I$ and $\gamma_{ij} = 1/|\mathcal{N}_i|, \forall j \in \mathcal{N}_i$.

Since the centralized Kalman filter, KCF and diffKF algorithms are inherently proposed for homogeneous sensor networks, here the simulations are performed over the sensor network consisting of only SNs with the same topology as shown in Fig. 4.2. As for DOCF-M algorithm, we consider two cases, namely, $|\mathcal{I}^c| = 0$ and $|\mathcal{I}^c| = 8$. In Fig. 4.6, we plot the comparison results regarding with MSD. Clearly, it shows that DOCF-M in the case of $|\mathcal{I}^c| = 0$ outperforms other distributed algorithms. Even if $|\mathcal{I}^c| = 8$ RNs are present in the network, MSD still remains at a satisfactory level, lower than that of Kalman-consensus filter (KCF) algorithm. Moreover, the result demonstrates that the optimal filter DOCF-M possesses improved estimation accuracy MSD compared with the non-optimal filter.

Additionally, we define an improvement factor (IF) of DOCF-M algorithm compared with KCF, diffKF and the non-optimal algorithms as

$$\text{IF} = \frac{\text{RMSE of the compared one} - \text{RMSE of DOCF-M}}{\text{RMSE of the compared one}}.$$

Fig. 4.6 Comparison of mean-square deviation (*MSD*) performance of different algorithms

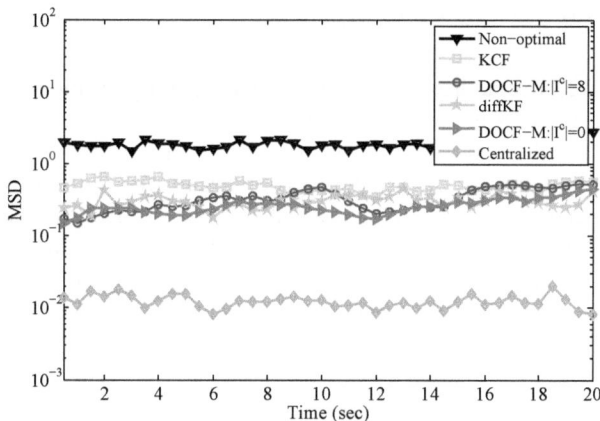

Table 4.2 summarizes RMSE of the four distributed algorithms and IF of DOCF-M compared to other algorithms. The result shows that DOCF-M in the case that there are no RNs, i.e., $|\mathcal{I}^c| = 0$, attains an improvement from a minimum of 5.0 % up to 62.96 % over KCF, diffKF and the non-optimal algorithms. In the case $|\mathcal{I}^c| = 8$, the optimal filter DOCF-M works much better than the nonoptimal filter with the improvement in RMSE being 56.51 %. This observation combined with that of Fig. 4.6 shows the advantage of optimal filter over nonoptimal one with respect to the estimation accuracy. In addition, the result implies that as for relay assisted WSN shown in Fig. 4.2, the RMSE is only a slight increase (i.e., 0.0624) compared with the homogeneous diffKF, where all SNs possess high processing abilities. This suggests that in the target tracking scenario the heterogeneous networks might be more appropriate choices than the homogeneous ones in terms of their reduction of overall cost with only little sacrifice of performance.

Table 4.2 RMSE and improvement factor (*IF*) of the proposed DOCF-M algorithm compared to some other distributed algorithms

		RMSE	IF
KCF	X: 0.4634		
	Y: 0.4604	0.6601	22.3%
diffKF	X: 0.3614		
	Y: 0.3944	0.5400	5.0%
Non-optimal: $|\mathcal{I}^c| = 8$	X: 0.9832		
	Y: 0.9749	1.3853	62.96%
DOCF-M: $|\mathcal{I}^c| = 8$	X: 0.4135		
	Y: 0.4103	0.6024	14.8%
DOCF-M: $|\mathcal{I}^c| = 0$	X: 0.3367		
	Y: 0.3723	0.5131	–

4.6 Summary

We have addressed the distributed tracking problem of a maneuvering target over relay assisted WSNs. A novel distributed optimal consensus filter is proposed to take the heterogeneity of node ability into account by solving the optimal control problems and optimization problems. Furthermore, we have investigated its convergence property. The theoretical analysis has been validated by simulation results that the estimation errors are exponentially bounded. The simulation results also suggest that relay assisted WSNs might be a more appropriate choice for the target tracking problem than the homogeneous one.

Chapter 5
Node Deployment for Distributed Consensus Estimation

This chapter deals with node deployment problem for distributed estimation of un-known signals in relay assisted WSNs. It is discovered that the network topology is closely related to the properties of the estimation algorithm. To satisfy the performance of the estimation algorithm, two node deployment algorithms are given. The first is concerned with the connectivity of network topology and the second illustrates a greedy approach to further optimize the network topology and the parameters of the estimation algorithms. Simulation results are provided to demonstrate the performance and effectiveness of the node deployment algorithms for solving distributed estimation problems in relay assisted WSNs.

5.1 Introduction

The node deployment is one of the fundamental problems for many applications in WSNs (see [112] and [113] for a comprehensive survey). A significant amount of studies of node deployment have focused on achieving network coverage and connectivity [114–117], prolonging the network lifetime [118] and guaranteeing fault tolerance of node failures [72, 119].

There are also studies on node deployment concerned with node localization, detection, and field estimation. In [120], Jourdan and Roy considered the case of deploying a sensor network that provides range measurements to a mobile agent for localization. The authors used the position error bound (PEB) as a measure of the quality of the node configuration and presented an iterative algorithm to place the nodes to minimize PEB. A related problem of node deployment for localization is the detection of targets entering guarded areas. Wettergren and Costa [121] studied the node deployment for surveillance of mobile targets. An optimization problem of maximizing the probability of successful detection against the targets is developed for placement of nodes. As for the node deployment for field estimation, it is usually referred to as how to estimate the quantity of interest at uncovered locations by using observations at locations with nodes [122]. There are two frequently used metrics: entropy and mutual information to measure the quality of the node configuration. In [123], Ko et al. studied the problem of selecting a subset of observations to minimize

© The Author(s) 2014   65
C. Chen et al., *Wireless Sensor Networks,*
SpringerBriefs in Computer Science, DOI 10.1007/978-3-319-12379-0_5

the uncertainty in spatial sampling networks, which is known as the subset selection problem. They showed that this problem is NP-hard. In this case, heuristic algorithms are widely used to solve these optimization problems. Different from [123], Krause et al. [124] considered the mutual information between the observed locations and those uncovered. The aim is to maximize this mutual information, which is proved to be NP-complete.

On the other hand, it has been mentioned in Chaps. 3 and 4 that not only the node configuration but also estimation algorithms should be carefully designed to fulfill detection and estimation problems over networks. In this chapter, we study the node deployment problem to ensure the performance of the distributed estimation in relay assisted WSNs. The node deployment problem in this chapter is similar to those considered in [72, 114, 119]. Typically, a common objective in such studies is to place a minimum number of RNs so that (1) the original or the whole network is k-connected ($k \geq 1$) and/or (2) coverage of the monitoring area is guaranteed. However what is investigated in this chapter is, to guarantee the distributed estimation performance together with connectivity. Thus, both the objective and methodologies in this chapter are quite different from the existing results. This chapter is based on [80]. The contributions are summarized as follows.

- Based on the distributed estimation algorithm proposed for relay assisted WSNs with two types of nodes of different computational capabilities, we properly construct a Markov chain with an absorbing state to discover that the RNs act as relays in the network. It is revealed that the placement of nodes plays an important role in the distributed estimation problems (Corollary 5.1).
- In order to ensure the performance of the proposed estimation algorithm, we propose a node deployment scheme to meet two requirements, namely, network connectivity and the number of neighbors of each node (Condition (5.10)). The scheme is divided into two phases. The first phase (Algorithm 2) deals with the connectivity requirement. In order to reduce the redundant nodes with the resulting network topology in the first phase, Algorithm 3 is proposed in the second stage to effectively delete some edges and adjust some parameters of the estimation algorithm to guarantee the estimation performance.

5.2 Problem Formulation

This section presents the estimation-oriented node deployment problem statement. It is noted that the models in this chapter can be considered as the scalar case of those in Chap. 3.

All the SNs can sense and observe the unknown signal $\zeta(t) \in \mathbb{R}$ with noisy and distorted measurements,

$$y_i(t) = b_i \zeta(t) + w_i(t), \tag{5.1}$$

where $b_i \geq 0$, and w_i are independent white Gaussian noises with zero mean and variances σ_i^2. This linear model (5.1) is widely used to describe several different

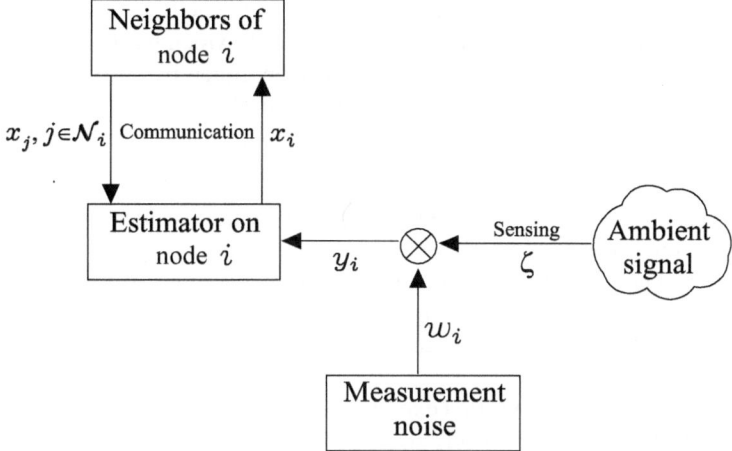

Fig. 5.1 A schematic representation of distributed estimation problem in *WSNs*

types of sensors, such as range bearing sensors and range-only sensors [125]. As for RNs, they do not make observations at all, i.e., they can only receive the local estimates from their neighbors and do some simple computations. They then forward the local estimates to their neighbors.

In this chapter, we model the communication topology $\mathcal{G}$ over which the nodes communicate as an undirected graph $\mathcal{G} = (\mathcal{V}, \mathcal{E})$ with $\mathcal{V}$ and $\mathcal{E}$ being the sets of nodes and symmetric communication links. Any two nodes can communicate with each other if and only if the distance between them is no longer than the communication radius r. We adopt the following algorithm on each node to account for computational heterogeneity of SNs and RNs. It can be regarded as the continuous-time version of the estimation algorithm used in Chap. 3. In detail, for SNs,

$$\dot{x}_i(t) = \beta_i x_i(t) + \alpha_i(y_i(t) - x_i(t)) + \frac{\sigma}{c_i} \sum_{j \in \mathcal{N}_i} a_{ij}[x_j(t) - x_i(t)], \ i \in \mathcal{I}, \quad (5.2)$$

and for RNs

$$x_i(t) = \sum_{j \in \mathcal{N}_i} \gamma_{ij} x_j(t), \ i \in \mathcal{I}^c, \quad (5.3)$$

where $\alpha_i > 0$, $\beta_i > 0$ are algorithm parameters, $\sigma > 0$ is the estimator gain, the other notations are defined in Chap. 3.

Note that (5.2) and (5.3) are different from the existing models [75, 126–130]. The fact that the estimator must take both filtering and consensus into consideration for all SNs motivates the new models (5.2) and (5.2). This distributed estimation problem in WSNs is schematically shown in Fig. 5.1 .

The ability of the relay assisted WSN to estimate unknown signals depends on the area coverage and connectivity of the network and the estimation algorithm. Area coverage refers to the union of the area monitored by the nodes. As for the estimation algorithm, convergence is the most important aspect, which has been discovered to be closely related with the network topology. Consequently, in order to guarantee the performance of the estimation, node deployment should be carefully considered.

Given the above discussions, the estimation-oriented node deployment problem can now be stated as follows:

Problem 5.1 (Estimation-orientated node deployment) Consider the scenario of estimating the unknown ambient signal $\zeta(t)$ by using a relay assisted WSN $\mathcal{G}$ composed of two types of nodes with different computational capabilities. We seek to place SNs over the monitoring area with coverage requirement and additional RNs to ensure the connectivity of the network of SNs and the performance of the estimation algorithms (5.2) and (5.3).

In the next two sections, we first explore the properties of the estimation algorithm proposed in this section and study its convergence analysis. We can then present a novel node deployment algorithm to ensure the performance of the estimation algorithm.

5.3 Analysis of the Proposed Estimation Algorithm

5.3.1 Properties of the Algorithm

We first introduce a graph $\tilde{\mathcal{G}}_s[\check{\mathcal{I}}]$, following four steps, which is a little different from the one proposed in Chap. 3 for directed graphs:

1. *Induction*: Derive the subgraph $\mathcal{G}[\check{\mathcal{I}}]$ of $\mathcal{G}$ induced by $\check{\mathcal{I}} := \tilde{\mathcal{I}} \cup \mathcal{I}^c$ via dropping the nodes outside $\check{\mathcal{I}}$ and the associated edges in graph $\mathcal{G}$.
2. *Deletion*: Delete the edges $(i, j) \in \mathcal{E}$ of which i, j are both in $\tilde{\mathcal{I}}$.
3. *Split*: Split each node $i \in \tilde{\mathcal{I}}$ into $|\mathcal{N}_i^R|$ different nodes possessing the same state, i.e., $x_{ij} = x_i$ for all $j = 1, 2, \ldots, |\mathcal{N}_i^R|$, such that each node i^j has only one neighbor in $\mathcal{I}^c$. Denote the set of all such nodes by $\tilde{\mathcal{I}}_s$ and the obtained graph by $\tilde{\mathcal{G}}_s[\check{\mathcal{I}}]$. In this way, all of the nodes in $\tilde{\mathcal{I}}_s$ are pendant ones.
4. *Partition*: Partition graph $\tilde{\mathcal{G}}_s[\check{\mathcal{I}}]$ into connected components (CCs), let us say, $\mathcal{G}_1 = (\mathcal{V}_1, \mathcal{E}_1), \ldots, \mathcal{G}_K = (\mathcal{V}_K, \mathcal{E}_K)$, where K is the number of CCs. Observe that if graph $\mathcal{G}$ is connected, then each $\mathcal{V}_k, k = 1, 2, \ldots, K$ contains at least one node of $\tilde{\mathcal{I}}_s$.

We use the example shown in Fig. 5.2a to illustrate the above four steps. Note that here $\tilde{\mathcal{I}} = \mathcal{I}$, then $\check{\mathcal{I}} = \mathcal{V}$. From Step 3, $\tilde{\mathcal{I}}_s$ can be expressed as $\tilde{\mathcal{I}}_s = \{1', 1'', 4', 4'', 4''', 7', 7''\}$, where $x_{1'} = x_{1''} = x_1$, $x_{4'} = x_{4''} = x_{4'''} = x_4$, and $x_{7'} = x_{7''} = x_7$. By operating the aforementioned steps, the resultant graph $\tilde{\mathcal{G}}_s[\check{\mathcal{I}}]$ with 3 CCs is shown in Fig. 5.2b. Since graph $\mathcal{G}$ is connected, i.e., any two nodes can reach from each other via multihop edges, each CC contains at least one node of $\tilde{\mathcal{I}}_s$, namely, $\{1''\} \subset$ CC $\mathcal{G}_1$, $\{1', 4''', 7''\} \subset$ CC $\mathcal{G}_2$ and $\{4', 4'', 7'\} \subset$ CC $\mathcal{G}_3$.

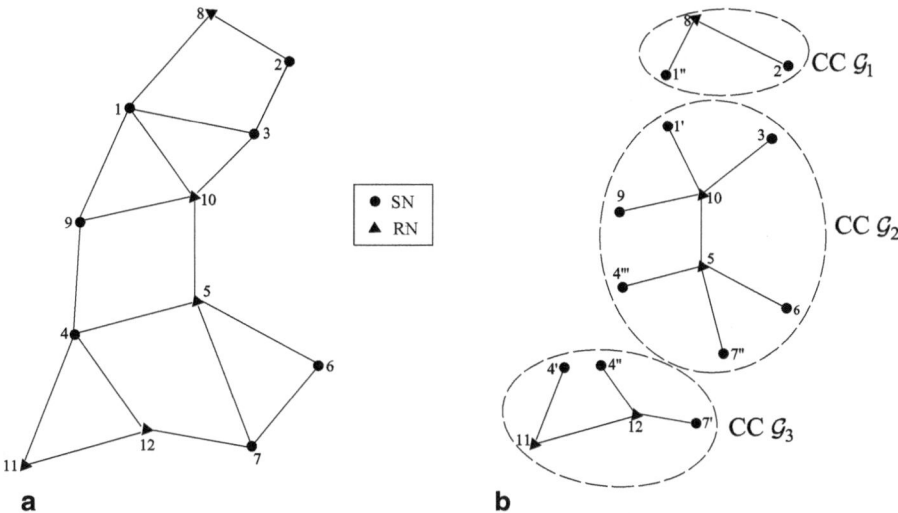

Fig. 5.2   a A sensor network $\mathcal{G}$ with 12 nodes consisting of 7 *SNs* and 5 *RNs*. **b** The corresponding graph $\tilde{\mathcal{G}}_s[\check{\mathcal{I}}]$ with 3 *CCs*, each one is enclosed in an ellipse

The graph $\tilde{\mathcal{G}}_s[\check{\mathcal{I}}]$ has the following important properties, whose proof follows a similar line of the directed counterpart Lemma 3.1 established in Chap. 3.

Proposition 5.1 *If graph $\mathcal{G}$ is connected, then the state of each node in $\mathcal{V}_k \backslash \tilde{\mathcal{I}}_s$ can be expressed as a convex combination of the states of nodes in $\mathcal{V}_k \cap \tilde{\mathcal{I}}$, $k = 1, 2, \ldots, K$, i. e.*

$$x_i = \sum_{j \in \mathcal{V}_k \cap \tilde{\mathcal{I}}} \tilde{\gamma}_{ij} x_j, \tag{5.4}$$

for all $i \in \mathcal{V}_k \backslash \tilde{\mathcal{I}}_s$, where $0 \le \tilde{\gamma}_{ij} \le 1$ and $\sum_{j \in \mathcal{V}_k \cap \tilde{\mathcal{I}}} \tilde{\gamma}_{ij} = 1$.

Proposition 5.1 reveals that the estimate of each RN is locally determined by those of SNs in the same CC. This is clear by carefully examining the models (5.2) and (5.3). This suggests that we only need to deal with SNs to solve the distributed estimation problem, once the weights γ_{ij}'s are given.

Now, substituting (5.4) to (5.2), we can obtain that, for node $i \in \mathcal{I}$,

$$\dot{x}_i = \alpha_i(y_i - x_i) + \beta_i x_i + \frac{\sigma}{c_i} \sum_{j \in \mathcal{N}_i^S} a_{ij}(x_j - x_i) + \frac{\sigma}{c_i} \sum_{j \in \mathcal{N}_i^R} a_{ij} \left(\sum_{k \in \tilde{\mathcal{I}}} \tilde{\gamma}_{jk} x_k - x_i \right)$$

$$= (-\alpha_i + \beta_i) x_i + \frac{\sigma}{c_i} \sum_{j \in \mathcal{N}_i^S} a_{ij}(x_j - x_i) + \alpha_i y_i + \frac{\sigma}{c_i} \sum_{j \in \mathcal{N}_i^R} a_{ij} \sum_{k \in \tilde{\mathcal{I}} \backslash \{i\}} \tilde{\gamma}_{jk}(x_k - x_i).$$

Without loss of generality, we assume that the SNs in $\mathcal{I}$ are the first M ones in $\mathcal{V}$. Concatenating x_i, y_i, $i \in \mathcal{I}$ in $x = [x_1, x_2, \ldots, x_M]^T$, $y = [y_1, y_2, \ldots, y_M]^T$, respectively, we can rewrite (5.2) into a compact form as follows

$$\dot{x}(t) = \left(\Phi - \sigma D_c^{-1} \hat{L} \right) x(t) + \Psi y(t), \tag{5.5}$$

where $\Phi = \mathrm{diag}\{\beta_1 - \alpha_1, \dots, \beta_M - \alpha_M\}$, $\Psi = \mathrm{diag}\{\alpha_1, \alpha_2, \dots, \alpha_M\}$, $D_c = \mathrm{diag}\{c_1, c_2, \dots, c_M\}$, and $\hat{L} = [\hat{L}_{ij}] \in \mathbb{R}^{M \times M}$ with entries

$$\hat{L}_{ij} = \begin{cases} \displaystyle\sum_{j \in \mathcal{N}_i^S} a_{ij} + \sum_{j \in \mathcal{N}_i^R} a_{ij} \sum_{k \in \hat{\mathcal{I}} \setminus \{i\}} \tilde{\gamma}_{jk}, & j = i, \\ -a_{ij} - \displaystyle\sum_{k \in \mathcal{N}_i^R} a_{ik} \tilde{\gamma}_{kj}, & j \neq i. \end{cases}$$

We have the following lemma, which is an immediate consequence of Theorem 3.2.

Lemma 5.1 *If graph $\mathcal{G}$ is connected, then zero is a simple eigenvalue of $\hat{L}$ and the corresponding left eigenvector $[\xi_1, \xi_2, \dots, \xi_M]^T$ is positive. Moreover, $\frac{1}{2}(\Xi \hat{L} + \hat{L}^T \Xi)$ is a positive semidefinite matrix.*

The above result will be utilized to analyze the convergence of the estimation algorithm in the next subsection.

5.3.2 Convergence Analysis on ε-Consensus

In this subsection, under some conditions, we shall establish that all the SNs can reach the ε-consensus asymptotically, which means under the model (5.2) and (5.3) each SN can approximately track the time-varying signal $\zeta(t)$ with the aid of RNs disregarding an ε-ball in the long run.

First, let us define the error variables $e_i(t) = x_i(t) - \zeta(t)$, $i \in \mathcal{I}$. Then we can collect $e_i(t)$'s into the following dynamics

$$\dot{e}(t) = \left(\Phi - \sigma D_c^{-1} \hat{L} \right) e(t) + \Psi y(t) + \zeta(t) \Phi \mathbf{1} - \dot{\zeta}(t) \mathbf{1}. \tag{5.6}$$

Now we are ready to state the main result about the performance of (5.2) and (5.3). The following theorem summarizes the convergence result for the noise-free case, i.e., $w_i = 0$, for all $i \in \mathcal{V}$.

Theorem 5.1 *Consider the connected network $\mathcal{G}$ composed of SNs and RNs. Suppose that there exist $v \geq 0, \mu \geq 0$ such that the signal $\zeta(t)$ satisfies $|\zeta(t)| \leq \mu$, $|\dot{\zeta}(t)| \leq v$ and*

$$\Theta := -\Xi D_c \Phi + \frac{\sigma}{2} \left(\Xi \hat{L} + \hat{L}^T \Xi \right) > 0, \tag{5.7}$$

then $e(t)$ asymptotically converges to a ε-ball with radius

$$\varepsilon = \frac{\varpi}{\lambda_{\min}(\Theta)} \sqrt{\frac{\bar{\eta}}{\underline{\eta}}},$$

as $t \to \infty$, *where* $\bar{\eta} = \max\{\xi_i c_i, i = 1, 2, \ldots, M\}$, $\underline{\eta} = \min\{\xi_i c_i, i = 1, 2, \ldots, M\}$ *and*

$$\varpi = \bar{\eta} \left(\nu \sqrt{M} + \mu \sqrt{\sum_{i=1}^{M} (\beta_i + \alpha_i (b_i - 1))^2} \right). \qquad (5.8)$$

Proof Since $\varXi D_c$ is positive definite from Lemma 5.1, it is natural to define the Lyapunov functional candidate as $V = \frac{1}{2} e^T \varXi D_c e$. The time derivative of V along the trajectory of the error dynamics (5.6) is given by

$$\dot{V} = \frac{1}{2} \left(e^T \varXi D_c \dot{e} + \dot{e}^T \varXi D_c e \right)$$

$$= e^T \left[\varXi D_c \varPhi - \frac{\sigma}{2} (\varXi \hat{L} + \hat{L}^T \varXi) \right] e + e^T \varXi D_c \left[\zeta(t)(b + \varPhi \mathbf{1}) - \dot{\zeta}(t) \mathbf{1} \right], \quad (5.9)$$

where $b = [b_1, b_2, \ldots, b_M]$. In view of the hypothesis (5.7), $\varXi D_c \varPhi - \frac{\sigma}{2} (\varXi \hat{L} + \hat{L}^T \varXi)$ is negative definite. Then employing the Schwarz inequality yields

$$\dot{V} \leq -\lambda_{\min}(\Theta) \|e\|^2 + \varpi \|e\|$$

$$= -\lambda_{\min}(\Theta) \left(\|e\| - \frac{\varpi}{2\lambda_{\min}(\Theta)} \right)^2 + \frac{\varpi^2}{4\lambda_{\min}(\Theta)},$$

where ϖ is defined in (5.8).

Let $\Omega_R = \{e : V(e) \leq R\}$ be the level-set of Lyapunov functional candidate V with $R = \frac{1}{2} \frac{\varpi^2 \bar{\eta}}{\lambda_{\min}^2(\Theta)}$, then for every $e \in \mathbb{R}^M \setminus \Omega_R$, it is easy to see that $\dot{V} < 0$. This means that if e is outside the set Ω_R, then there is an attractive force to pull it to Ω_R until it lies in Ω_R. As a result, for arbitrary $t_0 \in \mathbb{R}$, there exists $t_1 > t_0$ such that $e(t) \in \Omega_R, \forall t \geq t_1$, i.e., Ω_R is a trapping region for error dynamics (5.6). Consequently, for sufficiently large $t \in \mathbb{R}$, we can conclude that

$$\frac{1}{2} \underline{\eta} \|e\|^2 \leq V(e) \leq R = \frac{1}{2} \frac{\varpi^2 \bar{\eta}}{\lambda_{\min}^2(\Theta)}.$$

Therefore, $e(t)$ asymptotically converges to a ε-ball with the radius

$$\varepsilon = \frac{\varpi}{\lambda_{\min}(\Theta)} \sqrt{\frac{\bar{\eta}}{\underline{\eta}}}.$$

This completes the proof. □

The following corollary follows Theorem 5.1 to present a necessary condition. It reveals the relation between node configuration and distributed estimation performance in relay assisted WSNs.

Corollary 5.1 *For SNs, if* $\beta_i > \alpha_i$, *then it is necessary that the following condition*

$$\sum_{j \in \mathcal{N}_i} a_{ij} - \sum_{j \in \mathcal{N}_i^R} a_{ij} \tilde{\gamma}_{ji} > \frac{c_i}{\sigma} (\beta_i - \alpha_i) \qquad (5.10)$$

must hold in order to guarantee condition (5.7).

Proof The idea of the proof is based on the fact that the diagonal entries of a positive definite matrix must be positive as well.

In fact, the i-th diagonal entry of Θ is given by

$$\sigma \xi_i \hat{L}_{ii} + \xi_i(\alpha_i + \Lambda_i - \beta_i), \ i = 1, 2, \ldots, M.$$

It thus follows from (5.7) that the necessary condition is

$$\sigma \hat{L}_{ii} + \alpha_i + \Lambda_i - \beta_i > 0, \ i = 1, 2, \ldots, M,$$

since ξ_i is a positive scalar according to Lemma 5.1. Combined with the structure of $\hat{L}$, it infers that

$$\sigma \left(\sum_{j \in \mathcal{N}_i^S} a_{ij} + \sum_{j \in \mathcal{N}_i^R} a_{ij} \sum_{k \in \tilde{\mathcal{I}} \setminus \{i\}} \tilde{\gamma}_{jk} \right) > c_i(\beta_i - \alpha_i). \tag{5.11}$$

Note that $\tilde{\gamma}_{jk} = 0$, for all $j \in \mathcal{I}^c$ and $k \notin \tilde{\mathcal{I}}$ which is an immediate result from the construction of graph $\tilde{\mathcal{G}}_s[\tilde{\mathcal{I}}]$ in the subsection 5.3.1, then we have

$$\sum_{k \in \tilde{\mathcal{I}}} \tilde{\gamma}_{jk} = \sum_{k \in \mathcal{I}} \tilde{\gamma}_{jk}.$$

In consequence,

$$\sum_{j \in \mathcal{N}_i^R} a_{ij} \sum_{k \in \tilde{\mathcal{I}}} \tilde{\gamma}_{jk} = \sum_{j \in \mathcal{N}_i^R} a_{ij},$$

where use was made of Proposition 5.1. Hence substitution of the above identity into (5.11) yields

$$\sigma \left(\sum_{j \in \mathcal{N}_i} a_{ij} - \sum_{j \in \mathcal{N}_i^R} a_{ij} \tilde{\gamma}_{ji} \right) > c_i(\beta_i - \alpha_i).$$

This completes the proof. $\square$

Note that $0 \le \tilde{\gamma}_{ji} \le 1$, $\forall i \in \tilde{\mathcal{I}}$, $j \in \mathcal{I}^c$, then the following relations hold

$$\sum_{j \in \mathcal{N}_i} a_{ij} - \sum_{j \in \mathcal{N}_i^R} a_{ij} \tilde{\gamma}_{ji} = \sum_{j \in \mathcal{N}_i^S} a_{ij} + \sum_{j \in \mathcal{N}_i^R} a_{ij}(1 - \tilde{\gamma}_{ji}) \tag{5.12}$$

$$\ge \min_{j \in \mathcal{N}_i^S} a_{ij} |\mathcal{N}_i^S|. \tag{5.13}$$

As a result, if

$$|\mathcal{N}_i^S| > \frac{c_i}{\sigma \min_{j \in \mathcal{N}_i^S} a_{ij}} (\beta_i - \alpha_i) \tag{5.14}$$

is satisfied, then condition (5.10) must hold. This means that condition (5.14), and thus condition (5.10) to some extent, enforces a requirement on the number of neighbors of each SN for distributed estimation in relay assisted WSNs, i.e., the number of SN neighbors of each SN should be greater than a threshold related to its own confidence, estimator gain and the amplitude of signals sent from their neighbors. In other words, only the deployment of SNs matters for the performance of estimation algorithm (5.2) and (5.3). At first sight, this is doubtful, since the deployment of RNs seems to be irrelevant to the distributed estimation. However, after a second thought, this is reasonable by recalling that the SNs are more powerful than RNs. The estimation accuracy is primarily determined by the configuration of SNs. Therefore, more attention should be paid to the deployment of SNs, if the number of SNs is large enough.

On the other hand, we can not place arbitrarily many SNs to make condition (5.14) enforced in view of the overall cost of the network. Thus condition (5.14) might be much restrictive in some scenarios. Consequently, in the next section, we use condition (5.10) instead of (5.14) coupled with the connectivity requirement needed in Theorem 5.1 to present the node deployment algorithms.

5.4   Node Deployment for Both Connectivity and Distributed Estimation Performance

Theorem 5.1 establishes the convergence property of the distributed estimation algorithm presented in Sect. 5.2. In this section, we aim at placing SNs to satisfy the conditions needed in Theorem 5.1. In order to guarantee the desired performance of the estimation algorithm, two main requirements of the node configuration are: connectivity requirement and the number of neighbors requirement in (5.10). Intuitively, the second requirement can be easily solved by placing all the SNs together. Actually, this does not work in our scenario, because area coverage is also necessary to monitor the area and estimate the ambient signal $\zeta(t)$ in relay assisted WSNs. There is a trade-off between these two factors.

In order to deploy the SNs such that the requirement of condition (5.10) is satisfied, we can follow two steps: first, deploy the SNs, then deploy the RNs. The reason behind this handling is that the left-hand side of (5.10), i.e., $\sum_{j \in \mathcal{N}_i} a_{ij} - \sum_{j \in \mathcal{N}_i^R} a_{ij} \tilde{\gamma}_{ji}$ can be rewritten as $\sum_{j \in \mathcal{N}_i^S} a_{ij} + \sum_{j \in \mathcal{N}_i^R} a_{ij} (1 - \tilde{\gamma}_{ji})$. Note that the first term depends only on SNs, while the second term on RNs. This observation enables the two-step implementation of node deployment. After the deployment of SNs in the first step, check whether condition (5.10) is satisfied for each SN with $\beta_i > \alpha_i$. If not, then deploy some additional RNs until it is satisfied. This can always work, since $\sum_{j \in \mathcal{N}_i^R} a_{ij} (1 - \tilde{\gamma}_{ji}) \geq 0, \forall i \in \mathcal{I}$.

Based on the two-step implementation of deployment of SNs and RNs, in this chapter, we focus on the deployment of RNs, given that the SNs have been deployed to meet the requirement of coverage of the monitoring area. The algorithms in [114, 115, 117] can be used here to deploy the SNs in the first step.

Algorithm 2 is concerned with the connectivity requirement of the network of the SNs by placing some additional RNs. In Algorithm 2, we first compute a weighted undirected complete graph, and the weight corresponding to each edge (i, j) is the

---

Algorithm 2 RN deployment to ensure connectivity in relay assisted WSNs

---

Input: A set of M SNs denoted by $\mathscr{I} = \{1, 2, \ldots, M\}$, their coordinates $p_1, p_2, \ldots, p_M$, the communication range r of both SNs and RNs and positive constants $\alpha_i, \beta_i, i = 1, 2, \ldots, M$.
Output: A set of RNs and the graph $\mathscr{S}$.
  1: Construct an undirected complete graph $\mathscr{G}'' = (\mathscr{I}, \mathscr{E}'')$ and associate each edge $(i, j) \in \mathscr{E}''$
     with weight $d_{ij} := \|p_i - p_j\|$.
  2: Define a subgraph $\mathscr{S} = (\mathscr{I}, \mathscr{E}')$ initialized as $\mathscr{S} = (\mathscr{I}, \varnothing)$.
  3: **for** each edge $(i, j) \in \mathscr{E}''$ in increasing order of weight **do**
  4:     **if** $d_{ij} \leq r$ **then**
  5:         Put edge (i, j) into graph $\mathscr{S}$.
  6:     **end if**
  7:     **if** $\mathscr{S}$ is connected **then**
  8:         **break**
  9:     **else**
 10:         **if** $(\beta_i \leq \alpha_i$ and $\beta_j \leq \alpha_j)$ or $(\beta_i > \alpha_i$ and $\beta_j > \alpha_j)$ **then**
 11:             **if** $r < d_{ij} \leq 2r$ **then**
 12:                 Place a RN at some position with distance r from p_i and p_j on the perpendicular
                     bisector of line segment $[p_i, p_j]$.
 13:             **else if** $d_{ij} > 2r$ **then**
 14:                 Place two RNs at positions $p' = (1 - \kappa)p_i + \kappa p_j$ and $p'' = \kappa p_i + (1 - \kappa)p_j$, respectively, where $\kappa = \frac{r}{d_{ij}}$.
 15:                 Place $\tau := \lfloor \frac{d_{ij} - 2r}{r} \rfloor$ RNs on the line segment $[p', p'']$ and the k-th one is at
                     $(1 - k\kappa')p' + k\kappa'p''$, $k = 1, 2, \ldots, \tau$, where $\kappa' = \frac{r}{\|p' - p''\|}$.
 16:             **end if**
 17:         **else if** $\beta_i \leq \alpha_i$ and $\beta_j > \alpha_j$ **then**
 18:             **if** $d_{ij} > r$ **then**
 19:                 Place $\tau + 2$ RNs on the line segment $[p_i, p_j]$ and the k-th one is at $k\kappa p_i + (1 - k\kappa)p_j$, $k = 1, 2, \ldots, \tau + 2$.
 20:             **end if**
 21:         **else if** $\beta_i > \alpha_i$ and $\beta_j \leq \alpha_j$ **then**
 22:             **if** $d_{ij} > r$ **then**
 23:                 Place $\tau + 2$ RNs on the line segment $[p_i, p_j]$ and the k-th one is at $(1 - k\kappa)p_i + k\kappa p_j$, $k = 1, 2, \ldots, \tau + 2$.
 24:             **end if**
 25:         **end if**
 26:         Put edge (i, j) into graph $\mathscr{S}$.
 27:     **end if**
 28: **end for**

---

distance d_{ij}. Second, we place additional RNs on each edge with weight greater than r to enable the multihop communication between the two SNs associated with this edge. This process ends until the graph $\mathcal{S}$ is connected. During this process, note that for such SNs with $\beta_i > \alpha_i$, the additional RNs are placed with distance r from them. The purpose is to make the value of the second term of left-hand side of (5.10) as small as possible, since a_{ij} is inversely proportional to d_{ij}. This will be further discussed in Algorithm 3.

Algorithm 3 illustrates a greedy algorithm to adjust the network topology and parameters of the estimation algorithm. This algorithm is divided into two phases. The first phase is used to delete the redundant edges and the resulting isolated RNs

---

Algorithm 3 Network topology optimization and algorithm design for distributed estimation in relay assisted WSNs

---

Input: Graph $\mathscr{G} = (\mathscr{I} \cup \mathscr{I}', \mathscr{E})$, where $\mathscr{I}'$ is the set of placed RNs in Algorithm 2, coordinates of all SNs $p_1, p_2, \ldots$ and the algorithm parameters σ, α_i, β_i, c_i and $\tilde{\gamma}_{ij}$, $i = 1, 2, \ldots, M$.

1: Associate each edge (i, j) of graph $\mathscr{G}$ with weight d_{ij}.
2: **for** each edge $(i, j) \in \mathscr{I} \times \mathscr{I}'$ in increasing order of weight **do**
3:     **if** subgraph $(\mathscr{I} \cup \mathscr{I}', \mathscr{E} \setminus \{(i, j)\})$ is connected **then**
4:         Delete edge (i, j) from graph $\mathscr{G}$.
5:     **end if**
6: **end for**
7: **for** each edge $(i, j) \in \mathscr{I}' \times \mathscr{I}'$ in decreasing order of weight **do**
8:     **if** subgraph $(\mathscr{I} \cup \mathscr{I}', \mathscr{E} \setminus \{(i, j)\})$ is connected **then**
9:         Delete edge (i, j) from graph $\mathscr{G}$.
10:     **end if**
11: **end for**
12: Discard those pendant RNs from graph $\mathscr{G}$.
13: **for** each SN $i \in \mathscr{I}$ **do**
14:     Compute the set of neighbors $\mathscr{N}_i$ of SN i and weight a_{ij}, $\forall j \in \mathscr{N}_i$.
15:     **if** $\beta_i > \alpha_i$ and (5.10) is violated **then**
16:         Decrease the value of c_i such that (5.10) is satisfied.
17:     **end if**
18: **end for**

---

from graph S generated by Algorithm 2. But put it back if it is necessary to guarantee the connectivity of graph S. Those edges between closer SNs and RNs are discarded first for the same reason as in Algorithm 2. After the edge pruning, if condition (5.10) is still not satisfied for some SNs $i = 1, 2, \ldots, M$, then we decrease the value of c_i in the second phase until it is finally satisfied.

5.5 Simulation Study

In this section, we consider the estimation problem using a relay assisted WSN in 2-D plane. The performances of the proposed estimation algorithm and node deployment algorithms are evaluated by several simulations.

In the following simulations, we first randomly place 30 SNs in the monitoring area $[-40, 40] \times [-40, 40]$ as shown in Fig. 5.3 such that the distance between any two SNs is at least 2.5. This setup is meant to meet the requirement of coverage. Each SN can communicate with others within a disc of radius $r = 10$ and can only measure a noise version of the ambient signal $y_i(t) = \zeta(t) + w_i(t)$, where $\zeta(t) = 3 \sin(t) + \cos(2t)$ and $w_i(t)$ is zero-mean white noise with intensity $\sigma_i^2 = 0.125$, for all $i = 1, 2, \ldots, 30$. We apply Algorithm 2 and Algorithm 3 described in Sect. 5.4 to determine the positions of the additional RNs such that the whole network is connected and condition (5.10) is satisfied. We first use Algorithm 2 to

Fig. 5.3 Topology of a
random network with 30 SNs
with the distances between
them being at least 2.5

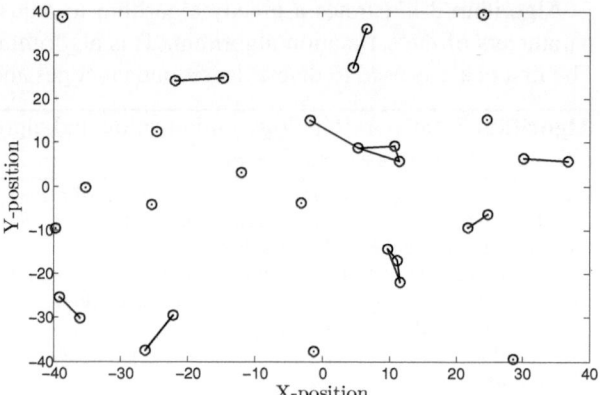

Fig. 5.4 The resulting relay
assisted WSN with 21 RNs
placed in the network shown
in Fig. 5.3 using Algorithm 2
and Algorithm 3

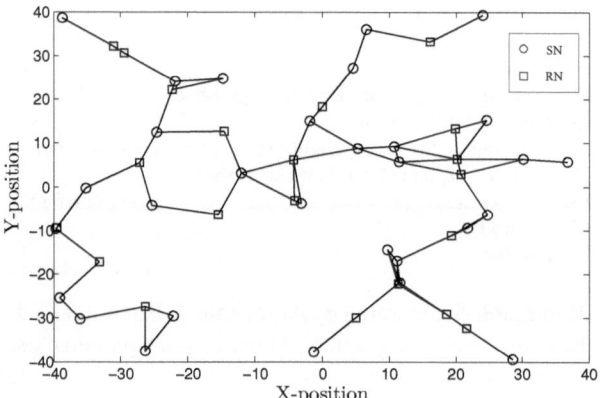

place the RNs until the network is connected. Twenty-two RNs are placed in this
phase. Then we employ Algorithm 3 to delete the redundant edges between RNs
and the corresponding RNs and adjust the parameters of the algorithm to enforce
condition (5.10). Here one RN and four edges are discarded with no influence on
the network connectivity. After this adjustment, we find that the sufficient condition
(5.7) is satisfied as well. Figure 5.4 depicts the corresponding relay assisted WSN,
where 21 RNs are placed. Although edges pruning is performed in Algorithm 3, it
can be easily seen that there is still edge redundancy in Fig. 5.4. This is normal in
greedy approaches, since greedy approaches (although efficient in most cases) may
be arbitrarily away from optimal solutions.

In the following simulations, we adopt a simple model for normalized path loss
without path fading $a_{ij} = \sqrt{1/(1 + d_{ij}^2)}$, where d_{ij} is the distance between SNs i and
j. The weight γ_{ij} is set to be

$$\gamma_{ij} = \frac{a_{ij}}{\sum_{j \in \mathcal{N}_i} a_{ij}}, \forall i \in \mathcal{I}^c, j \in \mathcal{V}.$$

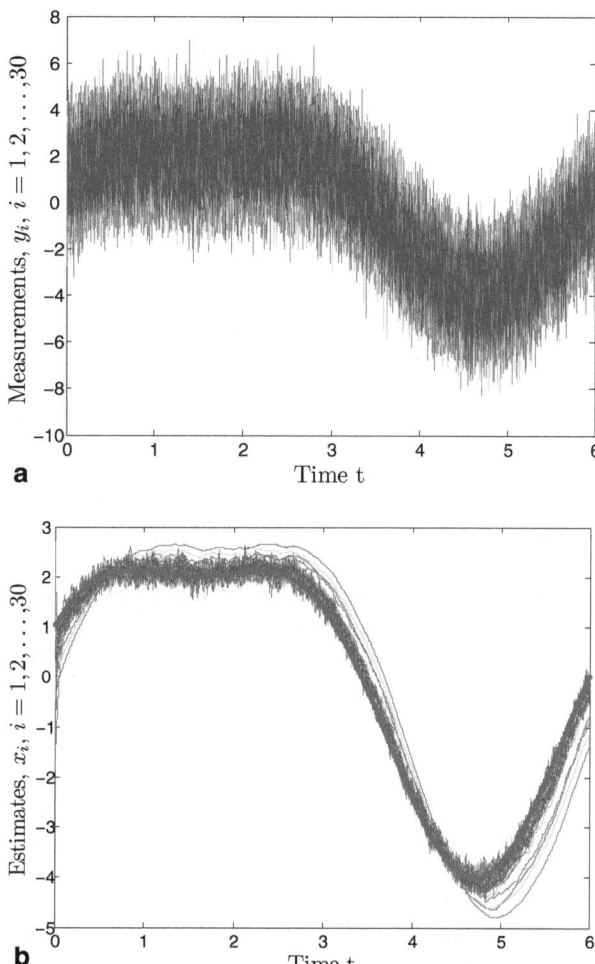

Fig. 5.5 **a** Measurements and **b** estimates of the ambient signal $\zeta(t)$ of all 30 *SNs*, where the *bold red curve* is the ambient signal $\zeta(t)$

Other parameters are selected as follows: $\sigma = 20$, $\alpha_i = 0.5$, $\beta_i = 1$, $c_i = 0.1$ if i is odd, and $\alpha_i = 40$, $\beta_i = 0.01$, $c_i = 0.05$ if i is even. The initial estimates of all SNs are randomly chosen from the interval $[-3, 3]$. All the results are presented by averaging 10 independent runs.

Figure 5.5 shows the measurements and estimates of ambient signal $\zeta(t)$. We can easily see that all the SNs are able to track the signal $\zeta(t)$ with an acceptable error bound determined by Theorem 5.1.

We use two metrics to measure the performance of the proposed estimation algorithm. The first one is the mean-square error over all SNs. $\text{MSE} = \frac{1}{30} \sum_{i=1}^{30} (x_i(t) - \zeta(t))^2$, which measures the estimation accuracy of the algorithm. And the second one characterizing the differences of the estimates among the SNs is the mean disagreement of the estimates defined by $\text{MSD} = \frac{1}{30} \sum_{i=1}^{30} (x_i(t) - \bar{x}(t))^2$, where $\bar{x} = \frac{1}{30} \sum_{i=1}^{30} x_i$. The simulation results of root-mean-square error (MSE) and

Fig. 5.6 *MSE* and *MSD* among all SNs

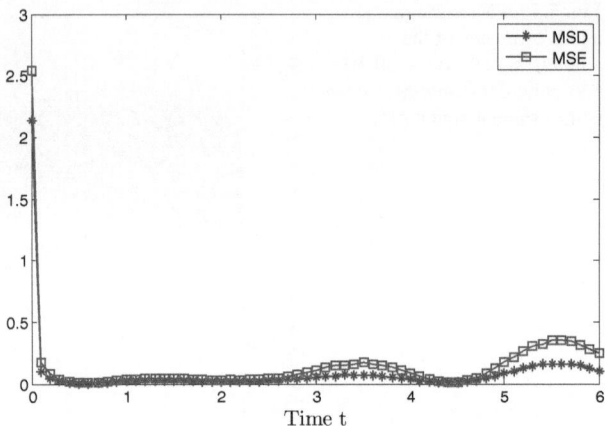

mean-square deviation (MSD) are shown in Fig. 5.6. It is observed that both MSD and MSE are enveloped by some curve, in other words, the proposed estimation algorithm works well, even if the environment is quiet noisy. Furthermore, small MSD means the proposed algorithm (5.2) possesses the ability of enabling all the SNs to cooperatively estimate the signal $\zeta(t)$ and reach agreement on the final estimate with a small error. Therefore, the proposed algorithm performs in a satisfactory manner. Finally, we compare the results with three different cases: $|\mathcal{I}| = 10, 20$, and 30 SNs, respectively, regarding the average of the root-mean-square error over all SNs

$$
\text{RMSE} = \frac{1}{|\mathcal{I}|} \sum_{i=1}^{|\mathcal{I}|} \sqrt{\frac{1}{T - [T/2] + 1} \sum_{k=[T/2]}^{T} (x_i(k) - \zeta(k))^2},
$$

where T is the total number of iterations and $[T]$ is the rounding integer.

Table 5.1 presents the comparison results of root-mean-square error (RMSE) in these three cases, where for each case, we randomly place $|\mathcal{I}|$ SNs separated by at least 2.5. The parameters stay the same for all three cases as adopted previously. The differences of RMSE are small which indicate that the estimation problem can be solved in all the three cases. And the RMSE is the smallest when the number of SNs is 20. This is because only the SNs make observations about the ambient signal ζ. If the number of SNs is small, few information could be sensed by the SNs. While if the number of SNs is large, more noise might be injected to the data about the signal ζ. Therefore, in both cases, lower RMSE is certainly not expected. This further indicates that the network topology itself matters more than the number of SNs for the distributed estimation in relay assisted WSNs. Extensive simulations and the optimal number of SNs, especially the theoretical analysis, for the distributed estimation based on the estimation algorithms (5.2) and (5.3) is one of our future works.

Table 5.1 Root-mean-square error (*RMSE*) of the abient signal $\zeta(t)$ for the proposed estimation algorithm (5.2) and (5.3) under the node deployment strategies Algorithm 2 and Algorithm 3

	$|\mathcal{I}| = 10$	$|\mathcal{I}| = 20$	$|\mathcal{I}| = 30$
RMSE	0.3525	0.2391	0.3222

5.6 Summary

The node deployment problem has been addressed for distributed estimation of unknown signals in relay assisted WSNs. Two classes of nodes with different processing abilities are taken into consideration: high quality SNs and low-end RNs. With the proposed estimation algorithm accounting for the computational heterogeneity, detailed properties of the proposed algorithm are analyzed. In order to ensure the performance of the distributed estimation, we present two node deployment algorithms to guarantee the connectivity and estimation performance. Simulation results have validated the theoretical analysis results.

Summary

Chapter 6
Conclusions and Future Work

In this chapter, we summarize the main results presented in this monograph and highlight future research directions.

6.1 Conclusions

In this monograph, we have investigated the network-wide estimation (and tracking) capability of wireless sensor network (WSNs) from the system perspective. This problem is of great importance since fundamental guidance on design and deployment of WSNs is vital for practical applications. Moreover, the large-scale of WSNs imposes distinguished challenges on systematic analysis and scalable algorithm design to satisfy fundamental estimation criteria, including network-wide convergence, unbiasedness, consistency, and optimality. We summarize the contributions of this monograph as follows.

- We have exploited the consensus-based estimation capability for a class of relay assisted sensor networks with asymmetric communication topology. By explicitly taking the heterogeneity of responsibilities between sensor nodes (for innovation based on local observations and cooperation between neighboring sensors) and relay nodes (for aggregation of data transmitted from neighboring nodes) into account, the distributed consensus-based unbiased estimation (DCUE) algorithm is proposed. With the algebraic graph theory in conjunction with Markov chain approach, the performance of asymptotic unbiasedness and consistency is analyzed for the DCUE algorithm. In addition, a quantitative bound of the convergence rate is established.
- For the same relay assisted networks but with symmetric communication topology, we investigate the problem of filter design for mobile target tracking over WSNs. Allowed with process noise of the target and observation noise of sensors, consensus-based distributed filters are designed for sensors to estimate the states (e.g., position and speed) of mobile target. The performance of unbiasedness and optimality are satisfied by optimal control technique. Moreover, the estimation errors are exponentially bounded.

© The Author(s) 2014
C. Chen et al., *Wireless Sensor Networks*,
SpringerBriefs in Computer Science, DOI 10.1007/978-3-319-12379-0_6

- It has been exploited on how to deploy sensor nodes and relay nodes to satisfied prescribed distributed estimation capability. A two-step algorithm is presented to meet the requirements of network connectivity and estimation performance. At the first step, the sensor nodes are roughly deployed to satisfy the connectivity requirement. The second step of the algorithm is to reduce redundant sensor nodes based on the resulting network topology analysis and adjust the estimation parameters to guarantee the asymptotic ϵ-consensus of estimation. It is discovered that, for a sensor node, the number of neighboring sensor nodes should be greater than a threshold related to its own confidence in the consensus process, estimation gain, and the estimates of its neighbors. This discovery then provides the guidance of deployment of relay nodes. The results in this book reveal from the system perspective that the network topology is closely related to the capability and efficiency of network-wide scalable distributed estimation.

6.2 Future Work

In this monograph, we focus on consensus-based estimation approach over WSNs. The increasing applications of cyber-physical systems (CPSs) witnesses the fact that the cooperative effort of sensor and actuator nodes can accomplish high-level situation awareness tasks with sensing, data processing, communication, and control. Our future work includes extensive validations based on real-world dataset, and further digging up the implication of network-wide estimation capability from the system perspective. We just list some of future research directions in this field.

- As far as distributed estimation is concerned, some possible directions remain to be further explored, such as the impact of network topology on the accuracy of estimation and tracking, further theoretical analysis in the presence of communication delay and so on. The distributed estimator/filter design with no prior knowledge about the process noise of the target is another future work.
- As for sensor deployment, some future directions include the investigation of the effects of link failures, communication noise on the performance of the algorithm, the deployment strategies of RNs aimed at better performance, etc.
- From the application perspective, more case studies of distributed estimation are desired to be presented with real data analysis and implementation in practical systems, such as situation awareness in industrial automation, stability enhancement in smart grid through state estimation, traffic estimation in intelligent transportation systems, and many other CPSs.

References

1. I. F. Akyildiz, W. Su, Y. Sankarasubramaniam, and E. Cayirci "A survey on sensor networks," *IEEE Communications Magazine*, vol. 40, no. 8, pp. 102–114, 2002.
2. J. Xiao, A. Ribeiro, Z. Luo, and G. B. Giannakis "Distributed compression-estimation using wireless sensor networks," *IEEE Signal Processing Magazine*, vol. 23, no. 4, pp. 27–41, 2006.
3. D. Bajovic, D. Jakovetic, J. M. Moura, J. Xavier, and B. Sinopoli "Large deviations performance of consensus + innovations distributed detection with non-gaussian observations," *IEEE Transactions on Signal Processing*, vol. 60, no. 11, pp. 5987–6002, 2012.
4. S. Kar, J. M. Moura, and K. Ramanan "Distributed parameter estimation in sensor networks: Nonlinear observation models and imperfect communication," *IEEE Transactions on Information Theory*, vol. 58, no. 6, pp. 3575–3605, 2012.
5. A. Y. Kibangou "Step-size sequence design for finite-time average consensus in secure wireless sensor networks," *Systems & Control Letters*, vol. 67, pp. 19–23, 2014.
6. L. Xiao and S. Boyd, "Fast linear iterations for distributed averaging," *Systems & Control Letters*, vol. 53, no. 1, pp. 65–78, 2004.
7. K. Avrachenkov, M. El Chamie, and G. Neglia "A local average consensus algorithm for wireless sensor networks," in *Proc. of The 2011 International Conference on Distributed Computing in Sensor Systems and Workshops (DCOSS'11)*, Barcelona, Spain, Jun.27–29 2011, pp. 1–6.
8. B. N. Oreshkin, M. J. Coates, and M. G. Rabbat "Optimization and analysis of distributed averaging with short node memory," *IEEE Transactions on Signal Processing*, vol. 58, no. 5, pp. 2850–2865, 2010.
9. J. M. Hendrickx, R. M. Jungers, A. Olshevsky, and G. Vankeerberghen "Graph diameter, eigenvalues, and minimum-time consensus," *Automatica*, vol. 50, no. 2, pp. 635–640, 2014.
10. A. Priolo, A. Gasparri, E. Montijano, and C. Sagues "A distributed algorithm for average consensus on strongly connected weighted digraphs," *Automatica*, vol. 50, no. 3, pp. 946–951, 2014.
11. S. Dasarathan, C. Tepedelenlioglu, M. Banavar, and A. Spanias "Non-linear distributed average consensus using bounded transmissions," *IEEE Transactions on Signal Processing*, vol. 61, no. 23, pp. 1–6, 2013.
12. Q. Zhang, B.-C. Wang, and J.-F. Zhang "Distributed dynamic consensus under quantized communication data," *International Journal of Robust and Nonlinear Control*, 2014, in press.
13. D. Thanou, E. Kokiopoulou, Y. Pu, and P. Frossard "Distributed average consensus with quantization refinement," *IEEE Transactions on Signal Processing*, vol. 61, no. 1, pp. 194–205, 2013.
14. L. Xiao, S. Boyd, and S.-J. Kim "Distributed average consensus with least-mean-square deviation," *Journal of Parallel and Distributed Computing*, vol. 67, no. 1, pp. 33–46, 2007.
15. R. Rajagopal and M. J. Wainwright, "Network-based consensus averaging with general noisy channels," *IEEE Transactions on Signal Processing*, vol. 59, no. 1, pp. 373–385, 2011.

© The Author(s) 2014
C. Chen et al., *Wireless Sensor Networks*,
SpringerBriefs in Computer Science, DOI 10.1007/978-3-319-12379-0

16. S. Barbarossa, L. Battisti, and S. Pescosolido "Distributed processing algorithms for wireless sensor networks having fast convergence and robustness against coupling noise," in *Proc. of The 10th IEEE International Symposium on Spread Spectrum Techniques and Applications (ISSSTA'08)*, Bologna, Italy, Aug.8–10 2008, pp. 1–6.

17. I. Matei, J. S. Baras, and C. Somarakis "Convergence results for the linear consensus problem under markovian random graphs," *SIAM Journal on Control and Optimization*, vol. 51, no. 2, pp. 1574–1591, 2013.

18. S. Kar and J. M. Moura, "Distributed consensus algorithms in sensor networks with imperfect communication: Link failures and channel noise," *IEEE Transactions on Signal Processing*, vol. 57, no. 1, pp. 355–369, 2009.

19. M. Kriegleder, R. Oung, and R. D'Andrea "Asynchronous implementation of a distributed average consensus algorithm," in *Proc. of The 2013 IEEE/RSJ Intelligent Robots and Systems (IROS'13)*, Tokyo, Japan, Nov.3–7 2013, pp. 1836–1841.

20. S. Li and Y. Guo, "Dynamic consensus estimation of weighted average on directed graphs," *International Journal of Systems Science*, no. ahead-of-print, pp. 1–15, 2013.

21. D. Li, Q. Liu, X. Wang, and Z. Lin "Consensus seeking over directed networks with limited information communication," *Automatica*, vol. 49, no. 2, pp. 610–618, 2013.

22. Y. Liu, D. W. Ho, and Z. Wang "A new framework for consensus for discrete-time directed networks of multi-agents with distributed delays," *International Journal of Control*, vol. 85, no. 11, pp. 1755–1765, 2012.

23. F. Pasqualetti, A. Bicchi, and F. Bullo "Consensus computation in unreliable networks: A system theoretic approach," *IEEE Transactions on Automatic Control*, vol. 57, no. 1, pp. 90–104, 2012.

24. S. Sundaram and C. N. Hadjicostis, "Distributed function calculation via linear iterative strategies in the presence of malicious agents," *IEEE Transactions on Automatic Control*, vol. 56, no. 7, pp. 1495–1508, 2011.

25. F. R. Yu, H. Tang, M. Huang, Z. Li, and P. C. Mason "Defense against spectrum sensing data falsification attacks in mobile ad hoc networks with cognitive radios," in *Military Communications Conference, 2009 (MILCOM'09)*. Boston, USA, Oct. 18–21, 2009, pp. 1–7.

26. S. Liu, H. Zhu, S. Li, X. Li, C. Chen, and X. Guan "An adaptive deviation-tolerant secure scheme for distributed cooperative spectrum sensing," in *Proc. of The 2012 IEEE Global Communications Conference (GLOBECOM'12)*, Anaheim, USA, Dec.3–7 2012, pp. 603–608.

27. Y. Mo and R. Murray, "Privacy preserving average consensus." *In Proc. of 53rd Conference on Decision and Control*, Dec. 15–17 2014.

28. J. He, P. Cheng, L. Shi, and J. Chen "SATS: secure average-consensus-based time synchronization in wireless sensor networks," *IEEE Transactions on Signal Processing*, vol. 61, no. 24, pp. 6387–6400, 2013.

29. H. Zhu, A. Cano, and G. B. Giannakis "Distributed consensus-based demodulation: algorithms and error analysis," *IEEE Transactions on Wireless Communications*, vol. 9, no. 6, pp. 2044–2054, 2010.

30. H. Paul, J. Fliege, and A. Dekorsy "In-network-processing: Distributed consensus-based linear estimation," *IEEE Communications Letters*, vol. 17, no. 1, pp. 59–62, 2013.

31. G. Xu, H. Paul, D. Wuebben, and A. Dekorsy "Fast distributed consensus-based estimation (fast-dice) for cooperative networks," in *Proc. of The 18th International ITG Workshop on Smart Antennas (WSA'14)*, Erlangen, Germany, Mar.12–13 2014, pp. 1–8.

32. I. D. Schizas, G. B. Giannakis, S. I. Roumeliotis, and A. Ribeiro "Consensus in Ad Hoc WSNs with noisy links—part ii: Distributed estimation and smoothing of random signals," *IEEE Transactions on Signal Processing*, vol. 56, no. 4, pp. 1650–1666, 2008.

33. H. Paul, B.-S. Shin, and A. Dekorsy "Distributed consensus-based linear estimation with erroneous links," in *Proc. of The 17th International ITG Workshop on Smart Antennas (WSA'13)*, Stuttgart, Germany, Mar.13–14 2013, pp. 1–5.

34. A. Bertrand and M. Moonen, "Consensus-based distributed total least squares estimation in ad hoc wireless sensor networks," *IEEE Transactions on Signal Processing*, vol. 59, no. 5, pp. 2320–2330, 2011.

35. A. Bertrand, "Low complexity distributed total least squares estimation in ad hoc sensor networks," *IEEE Transactions on Signal Processing*, vol. 60, no. 8, pp. 4321–4333, 2012.

36. A. Schmidt and J. M. Moura, "Distributed field reconstruction with model-robust basis pursuit," in *Proc. of The 2012 IEEE International Conference on Acoustics, Speech and Signal Processing (ICASSP'12)*, Kyoto, Japan, Mar.25–30 2012, pp. 2673–2676.

37. C. Ravazzi, S. Fosson, and E. Magli "Distributed iterative thresholding for l_0/l_1-regularized linear inverse problems," http://calvino.polito.it/ravazzi/publications/DITA.pdf, 2013.

38. V. Kekatos and G. B. Giannakis, "Distributed robust power system state estimation," *IEEE Transactions on Power Systems*, vol. 28, no. 2, pp. 1617–1626, 2013.

39. M. Ozay, I. Esnaola, F. T. Vural, S. R. Kulkarni, and H. V. Poor "Sparse attack construction and state estimation in the smart grid: Centralized and distributed models," *IEEE Journal on Selected Areas in Communications*, vol. 31, no. 7, pp. 1306–1318, 2013.

40. U. A. Khan, S. Kar, A. Jadbabaie, and J. M. Moura "On connectivity, observability, and stability in distributed estimation," in *Proc. of The 49th IEEE Conference on Decision and Control (CDC'10)*, Atlanta, GA, US, Dec.15–17 2010, pp. 6639–6644.

41. S. Kar and J. M. Moura, "Convergence rate analysis of distributed gossip (linear parameter) estimation: Fundamental limits and tradeoffs," *IEEE Journal of Selected Topics in Signal Processing*, vol. 5, no. 4, pp. 674–690, 2011.

42. Q. Zhang and J.-F. Zhang, "Distributed parameter estimation over unreliable networks with markovian switching topologies," *IEEE Transactions on Automatic Control*, vol. 57, no. 10, pp. 2545–2560, 2012.

43. H. Dong, Z. Wang, and H. Gao "Distributed filtering for a class of time-varying systems over sensor networks with quantization errors and successive packet dropouts," *IEEE Transactions on Signal Processing*, vol. 60, no. 6, pp. 3164–3173, 2012.

44. S. Kar and G. Hug, "Distributed robust economic dispatch in power systems: A consensus+ innovations approach," in *Proc. of The 2012 IEEE Power and Energy Society General Meeting*, San Diego, CA, US, Jul.22–26 2012, pp. 1–8.

45. L. Xie, D.-H. Choi, S. Kar, and H. V. Poor "Fully distributed state estimation for wide-area monitoring systems," *IEEE Transactions on Smart Grid*, vol. 3, no. 3, pp. 1154–1169, 2012.

46. S. Zhu, C. Chen, and X. Guan "Distributed optimal consensus filter for target tracking in heterogeneous sensor networks," in *Proc. of The 8th Asian Control Conference (ASCC'11)*, Kaohsiung, Taiwan, May.15–18 2011, pp. 806–811.

47. R. Olfati-Saber "Distributed Kalman filtering for sensor networks," in *Proc. of The 46th IEEE Conference on Decision and Control (CDC'07)*, Dec.12–14 2007, pp. 5492–5498.

48. R. Olfati Saber "Kalman-consensus filter: Optimality, stability, and performance," in *Proc. of The 48th IEEE Conference on Decision Control & The 28th Chinese Control Conference (CDC/CCC'09)*, Shanghai, China, Dec.15–18 2009, pp. 7036–7042.

49. R. Carli, A. Chiuso, L. Schenato, and S. Zampieri "Distributed Kalman filtering based on consensus strategies," *IEEE Journal on Selected Areas in Communications*, vol. 26, no. 4, pp. 622–633, 2008.

50. N. Ilic, M. Stankovillic, and S. Stankoviv "Adaptive consensus-based distributed target tracking in sensor networks with limited sensing range," *IEEE Transactions on Control Systems Technology*, vol. 22, no. 2, pp. 778–785, 2014.

51. D. W. Casbeer and R. Beard, "Distributed information filtering using consensus filters," in *Proc. of The 2009 American Control Conference (ACC'09)*, Jun.10–12 2009, pp. 1882–1887.

52. A. T. Kamal, J. A. Farrell, and A. K. R. Chowdhury "Information weighted consensus," in *Proc. of The IEEE 51st Annual Conference on Decision and Control (CDC'12)*, Maui, Hawaii, US, Dec.10–13 2012, pp. 2732–2737.

53. X. Li, H. Caimou, and H. Haoji "Distributed filter with consensus strategies for sensor networks," *Journal of Applied Mathematics*, vol. 2013, 2013.

54. P. Millán, L. Orihuela, C. Vivas, and F. R. Rubio "Distributed consensus-based estimation considering network induced delays and dropouts," *Automatica*, vol. 48, no. 10, pp. 2726–2729, 2012.
55. D. Y. Kim and M. Jeon, "Robust distributed Kalman filter for wireless sensor networks with uncertain communication channels," *Mathematical Problems in Engineering*, vol. 2012, 2012.
56. H. Song, L. Yu, and W.-A. Zhang "Distributed consensus-based Kalman filtering in sensor networks with quantised communications and random sensor failures," *IET Signal Processing*, vol. 8, no. 2, pp. 107–118, 2014.
57. W. Li, S. Zhu, C. Chen, and X. Guan "Distributed consensus filtering based on event-driven transmission for wireless sensor networks," in *Proc. of The 31st Chinese Control Conference (CCC'12)*, Hefei, China, Jul.25–27 2012, pp. 6588–6593.
58. P. Millán, L. Orihuela, I. Jurado, C. Vivas, and F. R. Rubio "Distributed estimation in networked systems under periodic and event-based communication policies," *International Journal of Systems Science*, no. ahead-of-print, pp. 1–13, 2013.
59. H. Long, Z. Qu, X. Fan, and S. Liu, "Distributed extended Kalman filter based on consensus filter for wireless sensor network," in *Proc. of The 10th World Congress on Intelligent Control and Automation (WCICA'12)*, Jul.6–8 2012, pp. 4315–4319.
60. D. Gu, J. Sun, Z. Hu, and H. Li "Consensus based distributed particle filter in sensor networks," in *Proc. of The 2008. International Conference on Information and Automation (ICIA'08)*, Jun.20–23 2008, pp. 302–307.
61. H. Liu, H. C. So, K. W. F. Chan, and W. K. Lui "Distributed particle filter for target tracking in sensor networks," *Progress In Electromagnetics Research C*, vol. 11, pp. 171–182, 2009.
62. J. Read, K. Achutegui, and J. Míguez "A distributed particle filter for nonlinear tracking in wireless sensor networks," *Signal Processing*, vol. 98, pp. 121–134, 2014.
63. L. Xiao, S. Boyd, and S. Lall "A scheme for robust distributed sensor fusion based on average consensus," in *Proc. of The 4th International Symposium on Information Processing in Sensor Networks (IPSN'05)*, Los Angeles, USA, Apr.15 2005, pp. 63–70.
64. I. D. Schizas, A. Ribeiro, and G. B. Giannakis "Consensus in Ad Hoc WSNs with noisy links-Part I: Distributed estimation of deterministic signals," *IEEE Transactions on Signal Processing*, vol. 56, no. 1, pp. 342–356, 2008.
65. A. K. Das and M. Mesbahi, "Distributed linear parameter estimation over wireless sensor networks," *IEEE Transactions on Aerospace and Electronic Systems*, vol. 45, no. 4, pp. 1293–1306, 2009.
66. A. T. Kamal, J. A. Farrell, and A. K. R. Chowdhury "Information weighted consensus filters and their application in distributed camera network," *IEEE Transactions on Automatic Control*, vol. 58, no. 12, pp. 3112–3125, 2013.
67. P. Gupta and P. R. Kumar, "The capacity of wireless networks," *IEEE Transactions on Information Theory*, vol. 46, no. 2, pp. 388–404, 2000.
68. M. Yarvis, N. Kushalnagar, H. Singh, A. Rangarajan, Y. Liu, and S. Singh "Exploiting heterogeneity in sensor networks," in *Proc. of The 24th IEEE International Conference on Computer Communications (INFOCOM'05)*, Miami, USA, Mar.13–17 2005, pp. 878–890.
69. X. Du, Y. Xiao, and F. Dai "Increasing network lifetime by balancing node energy consumption in heterogeneous sensor networks," *Wireless Communications and Mobile Computing*, vol. 8, pp. 125–136, 2008.
70. Q. Wang, K. Xu, G. Takahara, and H. Hassanein "Device placement for heterogeneous wireless sensor networks: Minimum cost with lifetime constraints," *IEEE Transactions on Wireless Communications*, vol. 6, no. 7, pp. 2444–2453, 2007.
71. X. Cheng, D. Z. Du, L. Wang, and B. Xu "Relay sensor placement in wireless sensor networks," *Wireless Networks*, vol. 14, no. 3, pp. 347–355, 2008.
72. X. Han, X. Cao, E. L. Lloyd, and C. C. Shen "Fault-tolerant relay node placement in heterogeneous wireless sensor networks," *IEEE Transactions on Mobile Computing*, vol. 9, no. 5, pp. 643–656, 2010.
73. S. S. Pereira and A. Pagès-Zamora, "Mean square convergence of consensus algorithms in random WSNs," *IEEE Transactions on Signal Processing*, vol. 58, no. 5, pp. 2866–2874, 2010.

74. S. Zhu, C. Chen, X. Ma, B. Yang, and X. Guan "Consensus based estimation over relay assisted sensor networks for situation monitoring, " IEEE Journal of Selected Topics in Signal Processing, to appear, 2014.
75. G. Scutari, S. Barbarossa, and L. Pescosolido "Distributed decision through self-synchronizing sensor networks in the presence of propagation delays and asymmetric channels," IEEE Transactions on Signal Processing, vol. 56, no. 4, pp. 1667–1684, 2008.
76. S. M. Kay, Fundamentals of Statistical Processing: Estimation Theory. Upper Saddle River, NJ: Prentice Hall, 1993.
77. M. E. Yildiz, A. Scaglione, and A. Ozdaglar "Asymmetric information diffusion via gossiping on static and dynamic networks," in Proc. of The 49th IEEE Conference on Decision and Control (CDC'10), Atlanta, USA, Dec.15–17 2010, pp. 7646–7472.
78. E. L. Lloyd and G. Xue, "Relay node placement in wireless sensor networks," IEEE Transactions on Computers, vol. 56, no. 1, pp. 134–138, 2007.
79. S. Misra, S. D. Hong, G. Xue, and J. Tang "Constrained relay node placement in wireless sensor networks: Formulation and approximations," IEEE/ACM Transactions on Networking, vol. 18, no. 2, pp. 434–448, 2010.
80. S. Zhu, C. Chen, and X. Guan "Sensor deployment for distributed estimation in heterogeneous wireless sensor networks," Ad Hoc & Sensor Wireless Networks, vol. 16, no. 4, pp. 297–322, 2012.
81. I. Maric, A. Goldsmith, and M. Mdard "Analog network coding in the high-SNR regime," in Proc. of The 2010 IEEE Wireless Network Coding Conference (WiNC'10), Boston, USA, Jun.21 2010, pp. 1–6.
82. C. W. Wu "On bounds of extremal eigenvalues of irreducible and m-reducible matricies," Linear Algebra and its Applications, vol. 402, pp. 29–45, 2005.
83. J. Kemeny and J. Snell, Eds., Finite Markov Chains. London: Springer-Verlag, 1976.
84. W. Ren and R. W. Beard, "Consensus seeking in multiagent systems under dynamically changing interaction topologies," IEEE Transactions on Automatic Control, vol. 50, no. 5, pp. 655–661, 2005.
85. R. A. Horn and C. R. Johnson, Matrix analysis. Cambridge: Cambridge University Press, 1985.
86. H. Zhang, F. L. Lewis, and Z. Qu "Lyapunov, adaptive, and optimal design techniques for cooperative systems on directed communication graphs," IEEE Transactions on Industrial Electronics, vol. 59, no. 7, pp. 3026–3041, 2012.
87. H. Robbins and D. Siegmund, "A convergence theorem for nonnegative almost supermartingales and some applications," in Optimizing Methods in Statistics, J. S. Rustagi, Ed., Academic Press, New York, 1971.
88. R. Olfati-Saber, J. A. Fax, and R. M. Murray "Consensus and cooperation in networked multi-agent systems," Proceedings of the IEEE, vol. 95, no. 1, pp. 215–233, 2007.
89. P. Barooah, J. P. Hespanha, and A. Swami "On the effect of asymmetric communication on distributed time synchronization," in Proc. of The 46th IEEE Conference on Decision and Control (CDC'07), New Orleans, USA, Dec.12–14 2007, pp. 5465–5471.
90. Y. Bar-Shalom, X. R. Li, and T. Kirubarajan, Estimation with Applications to Tracking and Navigation: Theory, Algorithms and Software. New York: John Wiley & Sons, 2001.
91. S. Oh, L. Schenato, P. Chen, and S. Sastry "Tracking and coordination of multiple agents using sensor networks: System design, algorithms and experiments," Proceedings of the IEEE, vol. 95, no. 1, pp. 234–254, 2007.
92. A. Ribeiro, I. D. Schizas, S. I. Roumeliotis, and G. B. Giannakis "Kalman filtering in wireless sensor networks," IEEE Control Systems Magazine, vol. 30, no. 2, pp. 66–86, 2010.
93. B. S. Y. Rao, H. F. Durrant-Whyte, and J. A. Sheen "A fully decentralized multi-sensor system for tracking and surveillance," The International Journal of Robotics Research, vol. 12, no. 1, pp. 20–44, 1993.
94. F. S. Cattivelli and A. H. Sayed, "Diffusion LMS strategies for distributed estimation," IEEE Transactions on Signal Processing, vol. 58, no. 3, pp. 1035–1048, 2010.

95. Z. Li, Z. Duan, G. Chen, and L. Huang "Consensus of multiagent systems and synchronization of complex networks: A unified viewpoint," *IEEE Transactions on Circuits and Systems—Part I: Fundamental Theory and Applications*, vol. 57, no. 1, pp. 213–224, 2010.

96. W. Yu, G. Chen, Z. Wang, and W. Yang "Distributed consensus filtering in sensor networks," *IEEE Transactions on Systems, Man, and Cybernetics—Part B: Cybernetics*, vol. 39, no. 6, pp. 1568–1577, 2009.

97. C. Huang, D. W. C. Ho, and J. Lu "Partial-information-based distributed filtering in two-targets tracking sensor networks," *IEEE Transactions on Circuits and Systems—Part I: Fundamental Theory and Applications*, vol. 59, no. 4, pp. 820–832, 2012.

98. S. Zhu, C. Chen, W. Li, B. Yang, and X. Guan "Distributed optimal consensus filter for target tracking in heterogeneous sensor networks," *IEEE Transactions on Cybernetics*, vol. 43, no. 6, pp. 1963–1976, 2013.

99. S. Zhu, C. Chen, X. Guan, and C. Long "An estimator model for distributed estimation in heterogenous wireless sensor networks," in *Proc. of The 2010 IEEE International Symposium on World of Wireless Mobile and Multimedia Networks (WoWMoM'10)*, Montreal, Canada, Jun.14–17 2010.

100. B. D. O. Anderson and J. B. Moore, *Optimal Filtering*. Englewood Cliffs, NJ: Prentice-Hall, 1979.

101. M. Athans and E. Tse, "A direct derivation of the optimal linear filter using the maximum principle," *IEEE Transactions on Automatic Control*, vol. AC-12, no. 6, pp. 690–698, 1967.

102. D. E. Kirk, *Optimal Control Theory: An Introduction*. Mineola, NY: Dover Publications, 2004.

103. L. Dieci and T. Eirola, "Positive definiteness in the numerical solution of Riccati differential equtions," *Numerische Mathematik*, vol. 67, no. 3, pp. 303-313, 1994.

104. Å Björck, *Numerical Methods for Least Squares Problems*. Philadelphia: SIAM, 1996.

105. A. H. Jazwinski, *Stochastic Processes and Filtering Theory*. New York: Academic Press, 1970.

106. K. Reif, S. Günther, E. Yaz, and R. Unbehauen "Stochastic stability of the continuous-time extended Kalman filter," *IEE Proceedings-Control Theory and Applications*, vol. 147, no. 1, pp. 45–52, 2000.

107. T.-J. Tarn and Y. Rasis, "Observers for nonlinear stochastic systems," *IEEE Transactions on Automatic Control*, vol. AC-21, no. 4, pp. 441–448, 1976.

108. X. Mao, *Stochastic Differential Equations and Applications*, 2nd ed. Chichester: Horwood, 2007.

109. J. S. Baras, A. Bensoussan, and M. R. James "Dynamic observers as asymptotic limits of recursive filters: Special cases," *SIAM Journal on Applied Mathematics*, vol. 48, no. 5, pp. 1147–1158, 1988.

110. R. A. Singer "Estimating optimal tracking filter performance for manned maneuvering targets," *IEEE Transactions on Aerospace and Electronic Systems*, vol. AES-6, no. 4, pp. 473–483, 1970.

111. X. R. Li and V. P. Jilkov , "Survey of maneuvering target tracking. Part I: Dynamic models," *IEEE Transactions on Aerospace and Electronic Systems*, vol. 39, no. 4, pp. 1333–1364, 2003.

112. X. Li, A. Nayak, D. Simplot-Ryl, and I. Stojmenovic "Sensor placement in sensor and actuator networks," *Wireless Sensor and Actuator Networks: Algorithms and Protocols for Scalable Coordination and Data Communication*, pp. 263–294, 2010.

113. M. Younis and K. Akkaya, "Strategies and techniques for node placement in wireless sensor networks: A survey," *Ad Hoc Networks*, vol. 6, pp. 621–655, 2008.

114. X. Bai, S. Kumar, D. Xuan, Z. Yun, and T. H. Lai "Deploying wireless sensors to achieve both coverage and connectivity," in *Proc. of The 7th ACM International Symposium on Mobile Ad Hoc Networking and Computing (MobiHoc'06)*, NewYork, USA, May.22–24 2006, pp. 131–142.

115. X. Li, H. Frey, N. Santoro, and I. Stojmenovic "Strictly localized sensor self-deployment for optimal focused coverage," *IEEE Transactions on Mobile Computing*, vol. 10, no. 11, pp. 1520–1533, 2011.

116. S. Misra, S. D. Hong, G. Xue, and J. Tang "Constrained relay node placement in wireless sensor networks to meet connectivity and survivability requirements," in *Proc. of The 27th IEEE Conference on Computer Communications (INFOCOM'08)*, Phoenix, AZ, Apr.13–18 2008.

117. H. Zhang and J. C. Hou, "Maintaining sensing coverage and connectivity in large sensor networks," *Ad Hoc & Sensor Wireless Networks*, vol. 1, no. 1–2, pp. 89–124, 2005.

118. K. Xu, H. Hassanein, G. Takahara, and Q. Wang "Relay node deployment strategies in heterogeneous wireless sensor networks," *IEEE Transactions on Mobile Computing*, vol. 9, no. 2, pp. 145–159, 2010.

119. J. L. Bredin, E. D. Demaine, M. T. Hajiaghayi, and D. Rus "Deploying sensor networks with guaranteed fault tolerance," *IEEE/ACM Transactions on Networking*, vol. 18, no. 1, pp. 216–228, 2010.

120. D. B. Jourdan and N. Roy, "Optimal sensor placement for agent localization," *ACM Transactions on Sensor Networks*, vol. 4, no. 3, p. 13, 2008.

121. T. A. Wettergren and R. Costa, "Optimal placement of distributed sensors against moving targets," *ACM Transactions on Sensor Networks*, vol. 5, no. 3, p. 26, 2009.

122. F. Bian, D. Kempe, and R. Govindan "Utility based sensor selection," in *Proc. of The 5th International Conference on Information Processing in Sensor Networks (IPSN'06)*, NewYork, USA, Apr.19–21 2006, pp. 11–18.

123. C.-W. Ko, J. Lee, and M. Queyranne "An exact algorithm for maximum entropy sampling," *Operations Research*, vol. 43, no. 4, pp. 684–691, 1995.

124. A. Krause, A. Singh, and C. Guestrin "Near-optimal sensor placements in gaussian processes: Theory, efficient algorithms and empirical studies," *The Journal of Machine Learning Research*, vol. 9, pp. 235–284, 2008.

125. X. R. Li and V. P. Jilkov, "A survey of maneuvering target tracking-Part III: Measurement models," in *Proc. of The 2001 SPIE Conference on Signal and Data Processing of Small Targets*, San Diego, USA, Jul.29 2001, pp. 423–446.

126. D. P. Spanos, R. Olfati-Saber, and R. M. Murray "Distributed sensor fusion using dynamic consensus," in *Proc. of The 16th International Federation of Automatic Control World Congress*, Prague, Czech, Jul.2–8 2005.

127. D. P. Spanos, R. O. Saber, and R. M. Murray "Dynamic consensus on mobile networks," in *Proc. of The 16th International Federation of Automatic Control World Congress*, Prague, Czech, Jul.2–8 2005.

128. R. Olfati-Saber and J. S. Shamma, "Consensus filters for sensor networks and distributed sensor fusion," in *Proc. of The 44th IEEE Conference on Decision and Control, and European Control Conference (CDC-ECC'05)*, Seville, Spain, Dec.15 2005, pp. 6698–6703.

129. S. Barbarossa and G. Scutari, "Decentralized maximum-likelihood estimation for sensor networks composed of nonlinearly coupled dynamical systems," *IEEE Transactions on Signal Processing*, vol. 55, no. 7, pp. 3456–3470, 2007.

130. K. M. Lynch, I. B. Schwartz, P. Yang, and R. A. Freeman, "Decentralized environmental modeling by mobile sensor networks," *IEEE Transactions on Robotics*, vol. 24, no. 3, pp. 710–724, 2008.